SAFE SCIENCE

Promoting a Culture of Safety in Academic Chemical Research

Committee on Establishing and Promoting a Culture of Safety
in Academic Laboratory Research

Board on Chemical Sciences and Technology
Division on Earth and Life Studies

Board on Human-Systems Integration
Division of Behavioral and Social Sciences and Education

NATIONAL RESEARCH COUNCIL
OF THE NATIONAL ACADEMIES

THE NATIONAL ACADEMIES PRESS
Washington, D.C.
www.nap.edu

THE NATIONAL ACADEMIES PRESS 500 Fifth Street, NW Washington, DC 20001

NOTICE: The project that is the subject of this report was approved by the Governing Board of the National Research Council, whose members are drawn from the councils of the National Academy of Sciences, the National Academy of Engineering, and the Institute of Medicine. The members of the committee responsible for the report were chosen for their special competences and with regard for appropriate balance.

This material is based upon work supported by the National Science Foundation under Grant No. CHE-1215772, the U.S. Department of Energy under Award Number DE-SC0007960, the National Institute of Standards and Technology under contract number SB1341-12-CQ-0036/13-100, ExxonMobil Chemical Company, E. I. du Pont de Nemours and Company, and the American Chemical Society.

International Standard Book Number-13: 978-0-309-30091-9
International Standard Book Number-10: 0-309-30091-6
Library of Congress Control Number: 2014949075

Additional copies of this report are available for sale from the National Academies Press, 500 Fifth Street, NW, Keck 360, Washington, DC 20001; (800) 624-6242 or (202) 334-3313; http://www.nap.edu/.

Printed in the United States of America

THE NATIONAL ACADEMIES

Advisers to the Nation on Science, Engineering, and Medicine

The **National Academy of Sciences** is a private, nonprofit, self-perpetuating society of distinguished scholars engaged in scientific and engineering research, dedicated to the furtherance of science and technology and to their use for the general welfare. Upon the authority of the charter granted to it by the Congress in 1863, the Academy has a mandate that requires it to advise the federal government on scientific and technical matters. Dr. Ralph J. Cicerone is president of the National Academy of Sciences.

The **National Academy of Engineering** was established in 1964, under the charter of the National Academy of Sciences, as a parallel organization of outstanding engineers. It is autonomous in its administration and in the selection of its members, sharing with the National Academy of Sciences the responsibility for advising the federal government. The National Academy of Engineering also sponsors engineering programs aimed at meeting national needs, encourages education and research, and recognizes the superior achievements of engineers. Dr. C. D. Mote, Jr., is president of the National Academy of Engineering.

The **Institute of Medicine** was established in 1970 by the National Academy of Sciences to secure the services of eminent members of appropriate professions in the examination of policy matters pertaining to the health of the public. The Institute acts under the responsibility given to the National Academy of Sciences by its congressional charter to be an adviser to the federal government and, upon its own initiative, to identify issues of medical care, research, and education. Dr. Victor J. Dzau is president of the Institute of Medicine.

The **National Research Council** was organized by the National Academy of Sciences in 1916 to associate the broad community of science and technology with the Academy's purposes of furthering knowledge and advising the federal government. Functioning in accordance with general policies determined by the Academy, the Council has become the principal operating agency of both the National Academy of Sciences and the National Academy of Engineering in providing services to the government, the public, and the scientific and engineering communities. The Council is administered jointly by both Academies and the Institute of Medicine. Dr. Ralph J. Cicerone and Dr. C. D. Mote, Jr., are chair and vice chair, respectively, of the National Research Council.

www.national-academies.org

COMMITTEE ON ESTABLISHING AND PROMOTING A CULTURE OF SAFETY IN ACADEMIC LABORATORY RESEARCH

Members

H. HOLDEN THORP (*Chair*), Washington University in St. Louis, MO
DAVID M. DEJOY (*Vice Chair*), University of Georgia, Athens
JOHN E. BERCAW, NAS, California Institute of Technology, Pasadena, CA
ROBERT G. BERGMAN, NAS, University of California, Berkeley
JOSEPH M. DEEB, ExxonMobil Corporation, Houston, TX
LAWRENCE M. GIBBS,* Stanford University, Stanford, CA
THEODORE GOODSON, III, University of Michigan, Ann Arbor
ANDREW S. IMADA, A. S. Imada and Associates, Carmichael, CA
KIMBERLY BEGLEY JESKIE, Oak Ridge National Laboratory, Oak Ridge, TN
BRADLEY L. PENTELUTE, Massachusetts Institute of Technology, Cambridge
KARLENE H. ROBERTS, University of California, Berkeley
JENNIFER M. SCHOMAKER, University of Wisconsin–Madison
ALICE M. YOUNG, Texas Tech University, Lubbock

National Research Council Staff

DOUGLAS FRIEDMAN, Study Director
TOBY WARDEN, Associate Director, Board on Human-Systems Integration (through January 3, 2014)
ELIZABETH FINKELMAN, Program Coordinator (as of August 12, 2013)
CARL GUSTAV ANDERSON, Research Associate (as of February 3, 2014)
AMANDA KHU, Administrative Assistant (through August 9, 2013)
NAWINA MATSHONA, Senior Program Assistant (as of October 21, 2013)
RACHEL YANCEY, Senior Program Assistant (through June 3, 2013)

* Resigned June 10, 2014.

BOARD ON CHEMICAL SCIENCES AND TECHNOLOGY

Members

TIMOTHY SWAGER (*Co-Chair*), NAS, Massachusetts Institute of Technology, Cambridge
DAVID WALT (*Co-Chair*), NAE, Tufts University, Medford, MA
HÉCTOR D. ABRUÑA, Cornell University, Ithaca, NY
JOEL C. BARRISH, Bristol-Myers Squibb, Princeton, NJ
MARK A. BARTEAU, NAE, University of Michigan, Ann Arbor
DAVID BEM, The Dow Chemical Company, Philadelphia, PA
ROBERT G. BERGMAN, NAS, University of California, Berkeley
JOAN BRENNECKE, NAE, University of Notre Dame, IN
HENRY E. BRYNDZA, E. I. du Pont de Nemours & Company, Wilmington, DE
MICHELLE V. BUCHANAN, Oak Ridge National Laboratory, Oak Ridge, TN
DAVID W. CHRISTIANSON, University of Pennsylvania, Philadelphia
RICHARD EISENBERG, NAS, University of Rochester, NY
JILL HRUBY, Sandia National Laboratories, Albuquerque, NM
FRANCES S. LIGLER, NAE, University of North Carolina, Chapel Hill and North Carolina State University, Raleigh
SANDER G. MILLS, Merck Research Laboratories (Ret.), Scotch Plains, NJ
JOSEPH B. POWELL, Shell, Houston, TX
ROBERT E. ROBERTS, Institute for Defense Analyses, Alexandria, VA
PETER J. ROSSKY, NAS, Rice University, Houston, TX
DARLENE SOLOMON, Agilent Technologies, Santa Clara, CA

National Research Council Staff

TERESA FRYBERGER, Director
DOUGLAS FRIEDMAN, Senior Program Officer
KATHRYN HUGHES, Senior Program Officer
CARL-GUSTAV ANDERSON, Research Associate
ELIZABETH FINKELMAN, Program Coordinator
NAWINA MATSHONA, Senior Program Assistant

Preface

While the hazards of academic chemical research have long been recognized, recent incidents prompted the National Research Council to ask whether there was another way to look at instilling stronger safety practices in chemical research. In particular, could the ideas and methodologies of safety culture from the industrial sector, including non-laboratory settings such as the airline industry, health care, and manufacturing, be brought in a more intentioned way to produce recommendations for making laboratory science safer? As such, a panel was formed consisting of university academic leadership and safety and health administrators, highly distinguished chemistry faculty members, and experts in the field of safety culture and human–systems integration.

The committee brought expertise and outlooks that had never been assembled similarly before. One member is a university provost who has been a dean, chemistry department chair, and chancellor during a time when numerous new regulations were being imposed on higher education, and thus understands the difficulty of achieving compliance and shifting culture. We had environmental health and safety officials from academia, industry, and the national labs who have years of experience in implementing safety regulations and encouraging safe science. We had senior, highly distinguished faculty members, who have led labs handling chemical hazards for decades and have seen the evolution in safety attitudes. We had young faculty just setting up their labs for the first time. And we had experts in safety culture and the behavioral sciences, who had been involved in numerous industries and had dealt with changes in practices that followed high-profile incidents of many different kinds.

The process of building a common language among this group of disparate perspectives was challenging, but worthwhile. Initially, it was not obvious to the group that social-behavioral heuristics and rubrics of safety culture could be applied to chemical research. Conversely, the specific practices of laboratory behavior and extraordinary autonomy afforded to chemical researchers when it comes to safety were new to the safety culture experts. We persevered in these conversations, came to common understandings, and achieved results that we believe are unusual and important.

The committee engaged a similarly wide group, ranging from young graduate students just beginning to work with chemical hazards to seasoned laboratory veterans. We talked to individuals from both highly resourced schools with large research budgets and operations as well as regional public universities and private liberal arts colleges that had only one person working in environmental health and safety. We talked to faculty members whose expertise varied from ultrafast laser spectroscopy to an anthropologist who studies power dynamics in academic laboratories.

For decades, laboratory incidents have resulted in new regulations. The committee upholds that compliance is important and that there is always room for better adherence to regulations, which make research safer. However, in writing our recommendations, we strove not to simply produce a list of new regulations. Rather, we hoped that our report would move chemical research beyond simple compliance to the adoption of a culture of safety in academic laboratories that transcends inspections, standard operating procedures, and chemical safety plans. A true safety culture represents a total commitment to achieving safety even in the absence of specific rules or other regulatory guidance. It means making safety an ongoing operational priority.

Our recommendations challenge many longstanding ideas about chemical research. Working long hours and late into the night are still seen as rites of passage in the development of scientists. Student desks for data analysis, writing, and eating still persist inside the laboratories. Principal investigators and visitors to the laboratory often feel that they do not need personal protective equipment if they are not handling any hazardous materials. From our work, we believe there is eagerness among young scientists and veterans alike to challenge these assumptions.

H. Holden Thorp, *Chair*
David M. DeJoy, *Vice Chair*
Douglas Friedman, *Study Director*
Committee on Establishing and Promoting
a Culture of Safety in Academic Research Laboratories

Acknowledgment of Reviewers

This report has been reviewed in draft form by individuals chosen for their diverse perspectives and technical expertise, in accordance with procedures approved by the National Research Council's Report Review Committee. The purpose of this independent review is to provide candid and critical comments that will assist the institution in making its published report as sound as possible and to ensure that the report meets institutional standards for objectivity, evidence, and responsiveness to the study charge. The review comments and draft manuscript remain confidential to protect the integrity of the deliberative process. We wish to thank the following individuals for their review of this report:

Donna Blackmond, NAE, *The Scripps Research Institute*
Dominick J. Casadonte, Jr., *Texas Tech University*
Sharon Clarke, *University of Manchester*
Kenneth Fivizzani, *Nalco Company (ret.)*
David A. Hofmann, *University of North Carolina, Chapel Hill*
Robin Izzo, *Princeton University*
Brian M. Kleiner, *Virginia Polytechnic Institute and State University*
David Korn, IOM, *Massachusetts General Hospital* and *Harvard Medical School*
Stephen R. Leone, NAS, *University of California, Berkeley*
William Tolman, *University of Minnesota*

Although the reviewers listed above have provided many constructive comments and suggestions, they were not asked to endorse the conclusions or recommendations, nor did they see the final draft of the report before its release. The review of this report was overseen by **Julia M. Phillips**, NAE, *Sandia National Laboratories*, and **Jeffrey J. Siirola**, NAE, *Eastman Chemical Company (ret.)*. Appointed by the National Research Council, they were responsible for making certain that an independent examination of this report was carried out in accordance with institutional procedures and that all review comments were carefully considered. Responsibility for the final content of this report rests entirely with the authoring committee and the institution.

Contents

Summary

Recent serious and sometimes fatal accidents in chemical research laboratories at U.S. universities have driven government agencies, professional societies, industries, and universities themselves to examine the culture of safety in research laboratories. These incidents have triggered a broader discussion of how serious incidents can be prevented in the future and how best to train researchers and emergency personnel to respond appropriately when incidents do occur. As the priority placed on safety increases, many institutions have expressed a desire to go beyond simple compliance with regulations to work toward fostering a strong, positive safety culture: affirming a constant commitment to safety throughout their institutions, while integrating safety as an essential element in the daily work of laboratory researchers (Box S-1).

The shift away from mere compliance and toward promoting a strong, positive safety culture has already yielded benefits in industries such as aviation and health care. However, the best approach to promote an improved safety culture within the academic research environment—with its unique goals, cultural dynamics, practices, and pressures—merits investigation. At the request of the study sponsors,[1] the National Research Council appointed a committee of experts in chemistry, human–systems integration, laboratory safety management, university administration,

[1] This study was supported by the National Science Foundation, the U.S. Department of Energy, ExxonMobil Chemical Company, E.I. du Pont de Nemours and Company, the American Chemical Society, and the National Institute of Standards and Technology.

BOX S-1
What Is Safety Culture?

Safety culture refers to an organization's shared values, assumptions, and beliefs specific to workplace safety or, more simply, the importance of safety within the organization *relative to other priorities.*

A strong, positive safety culture arises not because of a set of rules, but because of a commitment to safety throughout an organization. Such a culture supports the free exchange of safety information, emphasizes learning and improvement, and assigns greater importance to identifying and solving problems rather than placing blame. High importance is assigned to safety all the time, not just when it is convenient or does not threaten personal or institutional productivity goals.

and other fields to examine the culture of safety in academic institutions and recommend ways to improve their overall safety performance. While this report is focused primarily on academic chemistry laboratories, there are a wide variety of environments both in and outside academia that may benefit from the recommendations made herein. The full statement of task can be found in Box 1-1 (Chapter 1).

During the course of its study, the committee heard from researchers, faculty, and others involved in chemical research, made site visits to academic labs, examined research literature on safety culture in other industries, and drew upon their own expertise to arrive at a series of findings and conclusions (see Chapter 5) about current safety culture and practices in academia. In addition, the committee recommends a series of actions that universities should take to build and sustain a strong, positive safety culture in their laboratories, with the ultimate goal of protecting the lives and health of the researchers who work in them.

Interest in promoting safety in academic research laboratories has grown in recent years, following high-profile incidents in which researchers were injured or killed. Many colleges and universities are interested in fostering a safety culture that goes beyond compliance with regulations: affirming a constant, institution-wide commitment to safety and integrating safety as an essential element in the daily work of researchers.

SAFETY CULTURE IN CHEMICAL RESEARCH

It is important to recognize that while fostering a strong, positive safety culture in academic research laboratories can reduce the risk of incidents and injuries, it cannot eliminate that risk entirely. The major objective of chemistry research endeavors, like all research, is to expand knowledge, and this pursuit entails experiments that may involve hazardous substances and new reactions, the nature and magnitude of which cannot always be predicted. The objective of establishing a strong, positive safety culture in a research setting is not to remove all risk—an impossible task—but to identify and mitigate hazards that are foreseeable, employ general precautions that help protect against unforeseeable hazards, and ensure the capacity to respond to incidents in ways that minimize harm.

An ideal laboratory safety culture ensures that all researchers who set foot in an academic laboratory, from inexperienced students to senior principal investigators (PIs), understand that they are entering a research environment that requires special precautions. Researchers are aware of the hazards posed by the materials with which they and other labs are working, and they are prepared to take rapid and appropriate measures to protect themselves and their co-workers, especially in the case of unexpected events.

A strong, positive safety culture encourages all laboratory workers to place the highest priority on best practices and to raise concerns to colleagues and supervisors, including principal investigators, when they identify or are concerned about potential safety problems. It is not enough to provide safe equipment, systems, and procedures if the culture of the organization does not encourage and support working safely in the research laboratory.

The specialized and insular structure and hierarchical nature of academic research can pose challenges to the development of a strong, positive safety culture. Principal investigators operate autonomously, exercising significant authority over the research and the research personnel in their individual laboratories, and in some cases may regard good safety practices, such as inspections by outsiders or following established safety procedures, as a barrier to research progress and a violation of their academic freedom. The very character of academic research and its pursuit of new knowledge engenders an entrepreneurial spirit, an aspect of which can be resistant to central dictates or "one-size-fits-all" mandates. Meanwhile, graduate students, postdoctoral fellows, and other research staff are dependent, financially and educationally, upon their principal investigators' grants and research projects. Concern about their future and the impact of their attitudes on their budding careers may make them reluctant to raise safety questions or concerns.

Overcoming these challenges and building cultures that prioritize safety will require responsibility and action from everyone involved in the research enterprise. Institutional leaders need to rethink how they deploy resources, organize reporting relationships, and structure incentives for promoting safety. Principal investigators will need to take responsibility for supporting and fostering safety culture in their laboratories, which includes taking proactive steps to counter the dynamics of the power differential that may inhibit laboratory researchers from raising or elevating safety concerns. Each individual researcher, whose safety is at stake, should play a leadership role in developing and sustaining strong safety culture in the laboratories where they work. Finally, environmental health and safety personnel should work collaboratively with all of these parties, assisting their efforts to establish a strong, positive safety culture.

RECOMMENDATIONS

Institution-Wide Dynamics and Resources

The broad institutional setting in which research takes place can strongly influence whether university laboratories develop and sustain a strong, positive safety culture. Specifically, the level of importance attached to safety by university leadership, the way these leaders promote safety as a core institutional value, the way they direct resources, and the structure of incentives and reporting relationships they support all affect the degree of priority given to safety practices.

Recommendation 1: The president and other institutional leaders must actively demonstrate that safety is a core value of the institution and show an ongoing commitment to it.

Recommendation 2: The provost or chief academic officer, in collaboration with faculty governance, should incorporate fostering a strong, positive safety culture as an element in the criteria for promotion, tenure, and salary decisions for faculty.

Recommendation 3: All institutions face a challenge of limited resources. Within this constraint, institutional head(s) of research and department chairs should consider the resources they have available for safety when considering or designing programs, and identify types of research that can be done safely with available and projected resources and infrastructure.

Recommendation 4: University presidents and chancellors should establish policy and deploy resources to maximize a strong, positive safety culture. Each institution should have a comprehensive risk management plan for laboratory safety that addresses prevention, mitigation, and emergency response.

These leaders should develop risk management plans and mechanisms with input from faculty, students, environmental health and safety staff, and administrative stakeholders and ensure that other university leaders, including provosts, vice presidents for research, deans, chief administrative officers, and department chairs, do so as well.

Research Group Dynamics

Many research groups have differential power dynamics, which, if not appropriately addressed, can work against the development of a strong, positive safety culture. Department chairs and principal investigators should take steps to change these dynamics, creating mechanisms that empower laboratory researchers to communicate freely about safety and take an active role in establishing and promoting a strong, positive safety culture and in sustaining a safe research enterprise.

Recommendation 5: Department chairs and principal investigators should make greater use of teams, groups, and other engagement strategies and institutional support organizations (e.g., environmental health and safety, facilities), to establish and promote a strong, positive, safety culture.

Recommendation 6: Department chairs should provide a mechanism for creating a robust safety collaboration between researchers, principal investigators, and environmental health and safety personnel.

Data, Hazard Identification, and Analysis

In addition to improving the organizational dynamics that drive safety practice, laboratories have a need for data and to conduct analyses that will help them identify and mitigate hazards. Traditionally, safety performance has been measured using lagging or after-the-fact indicators, such as numbers of accidents and lost-time injuries. To change behavior and culture before an incident occurs, organizations may take advantage of *leading indicators*: before-the-fact data that can help identify risks and vulnerabilities ahead of time. One key approach to identify hazards before they cause harm is to report and collect data on near misses. Another way to identify hazards is to conduct hazard analysis, a process to assess risks and their consequences and ensure that they are mitigated or eliminated before any lab work is initiated.

Recommendation 7: Organizations should incorporate non-punitive incident and near-miss reporting as part of their safety cultures. The American Chemical Society, Association of American Universities, Association of Public and Land-grant Universities, and American Council on Education should work together to establish and maintain an anonymous reporting system, building on industry efforts, for centralizing the collection of information about and lessons learned from incidents and near misses in academic laboratories, and linking these data to the scientific literature. Department chairs and university leadership should incorporate the use of this system into their safety planning. Principal investigators should require their students to utilize this system.

Recommendation 8: The researcher and principal investigator should incorporate hazard analysis into laboratory notebooks prior to experiments, integrate hazard analysis into the research process, and ensure that it is specific to the laboratory and research topic area.

Training and Learning

Training in safety practices—both initial training and ongoing mentoring and support—is an essential element in developing and sustaining a strong, positive safety culture. This is particularly important with researchers in academic labs, who are often relatively young and have limited experience. Entering (and even experienced) students may not know how to assess the risks of what they are doing, how to assess changes in risks if they change a key experimental parameter, or how to keep a small error from causing major problems. Moreover, they may not realize that a process they used in the past without apparent incident was out of the ordinary or dangerous.

Recommendation 9: Department leaders and principal investigators, in partnership with environmental health and safety personnel, should develop and implement actions and activities to complement initial, ongoing, and periodic refresher training. This training should ensure understanding and the ability to execute proper protective measures to mitigate potential hazards and associated risks.

Actions for Key Stakeholders

As mentioned previously, everyone in the research enterprise has an important and individual role to play in establishing and promoting a strong, positive safety culture.

Presidents, chancellors, and provosts should discuss safety frequently and publicly and demonstrate through their actions that safety is a core value of the institution. They should deploy university resources in ways that support safety and reduce existing disincentives to safety practice—for example, by paying for personal protective equipment and hazardous waste disposal, so that PIs do not have to pay for such measures out of grant funding. Each institution should have a comprehensive risk management plan for laboratory safety that addresses prevention, mitigation, and emergency response. In addition, provosts should work with faculty governance to incorporate efforts to foster a strong, positive safety culture as an element in the criteria for promotion, tenure, and salary decisions for faculty.

Vice presidents for research and deans of schools and colleges should, in addition to deploying funds in ways that support safety, ensure that the lines of research undertaken by the institution are ones it has the capacity to perform safely. They can make certain that everyone involved in the research enterprise knows their role and responsibilities in supporting safety. They can develop reporting structures that support safety culture; an example would be for senior environmental health and safety (EHS) officials to report through the senior research management programs, typically at the vice president level or higher—a structure that may better integrate safety management into overall research management.

PIs and department chairs have responsibility for establishing strong safety culture in the laboratories they oversee. They should set an example by using safe practices and personal protective equipment, and they should ensure that researchers are properly trained in safety before they undertake any work. They should also take steps to counter the power dynamics that may make researchers—whose academic future largely depends on their PI—reluctant to raise safety concerns and questions. For example, they should encourage open dialogue about safety concerns among researchers in their labs, and establish regular times—such as "safety moments" at the beginning of lab meetings—where concerns can be raised. Establishing ongoing measures to support safety, such as unannounced walk-through inspections and non-punitive reporting systems for near misses, is also important. Department chairs, meanwhile, should work to build strong and cooperative relationships between their departments and EHS.

EHS professionals should partner with administrators, faculty, and researchers to go beyond compliance and establish a strong, positive safety culture. They should reach out to these groups as they undertake these actions, offering collaboration and support. These professionals have an important role and the interactions between them and the rest of the research community is an important aspect of a strong, positive safety culture.

Researchers have responsibility for supporting safety culture in the labs where they work, and have the most at stake in doing so. Some of the strongest safety cultures are ones where researchers have taken leadership roles. Researchers should be encouraged to take such roles—by serving on safety committees, for example, and by taking part in non-punitive, walk-through inspections of other labs. The institution, meanwhile, must provide researchers with the equipment, training, systems, and cultural support they need to work safely.

1

Introduction

The publicity surrounding recent incidents in university research laboratories continues to draw attention to the importance of promoting safety within academic laboratory settings. In addition to drawing significant attention to laboratory safety, these incidents have evoked a broad range of institutional responses. At the request of the study sponsors,[1] the National Research Council appointed a committee of experts to examine laboratory safety in academic and non-industrial chemical research settings and to provide recommendations, grounded in insights from behavioral science, on how to improve the overall safety performance of such laboratories (Box 1-1).

MOTIVATION AND BACKGROUND

Serious and sometimes fatal accidents in chemistry research laboratories at universities have driven government agencies and professional societies to engage in renewed efforts to examine safety in university labs. Investigations from recent, highly publicized incidents, including those occurring at UCLA in 2008 and Texas Tech in 2010, identified issues of preparedness, proper training, and adherence to laboratory safety protocols as precursors to the incidents that transpired. Sometimes, though,

[1] This study was supported by the National Science Foundation, the U.S. Department of Energy, ExxonMobil Chemical Company, E.I. du Pont de Nemours and Company, the American Chemical Society, and the National Institute of Standards and Technology.

BOX 1-1
Statement of Task

The National Research Council, through its Board on Chemical Science and Technology and Board on Human Systems Integration, will examine laboratory safety in chemical research in non-industrial settings. It will compare practices and attitudes in these settings with knowledge about promoting safe practices from the behavioral science literature. It will make recommendations for systems and practices that would improve the safety of chemistry research laboratories specifically and other non-industrial research laboratories more generally. It will

- Describe the current hierarchy of actors responsible for laboratory safety in U.S. education and in national laboratories. Identify the strengths and shortcomings of these hierarchies and how they impact the development of a culture of safety in academic research laboratories.
- Examine knowledge from the behavioral sciences and experience with safety systems from other sectors (such as industrial research facilities, nuclear energy, aviation, and medical) for key attributes of successful safety systems and cultures. Use this to draw lessons that could be applied in non-industrial laboratory research.
- Provide guidance on systems (such as training and reporting) that might be established, maintained, and utilized to raise the overall safety performance of U.S. chemistry research laboratories.
- Determine key actors required to achieve broad implementation of improved safety performance in research laboratories, especially in the U.S. higher educational system, and provide guidance on their roles and how they might be effectively engaged in improving safe laboratory practice.

The resulting findings and conclusions will be disseminated broadly to key actors in non-industrial laboratory safety.

even when carried out by researchers with extensive training and prudent behavior, standard safety precautions can fail, as tragically exemplified by the 1997 death of Karen Wetterhahn, a respected chemistry professor from Dartmouth College.

Dartmouth Incident

Karen Wetterhahn, a specialist in metal toxicology, was a professor of chemistry at Dartmouth College and founding director of the university's Toxic Metals Research Program. In August 1996, while transferring dimethylmercury between containers, Wetterhahn dropped one to several drops of the compound onto her left, gloved hand.[2] During the transfer, Wetterhahn observed the standard safety protocol at the time, conducting the transfer in a fume hood, wearing eye goggles, and disposable latex gloves. Wetterhahn thought nothing of the minor spill. When she was done, she cleaned her equipment, removed her gloves, and washed her hands. Roughly five months later, Wetterhahn began experiencing difficulty seeing, speaking, hearing, and walking. Upon medical examination, Wetterhahn was diagnosed with acute mercury toxicity due to exposure to dimethylmercury. Despite aggressive chelation therapy, her condition continued to deteriorate, and in February 1997, Wetterhahn went into a coma. She died on June 8, 1997, only ten months after the initial exposure.[3]

The unsettling characteristic of this incident is that Wetterhahn carried out the dimethylmercury transfer appropriately and safely to the best of anyone's knowledge at the time. Notably, the Material Safety Data Sheets (MSDS) for dimethylmercury recommended the use of rubber, neoprene, or otherwise "chemically impervious gloves" when handling the compound. The MSDS offered no additional detail on the subject. Following Wetterhahn's death, permeation testing of disposable latex gloves revealed that dimethylmercury permeates latex, PVC, and neoprene almost immediately upon contact.[4] Acknowledging the great risk associated with handling dimethylmercury as well as its lethal properties, OSHA amended its safety guidelines for the compound, discouraging its further use, unless absolutely necessary. In OSHA's memorandum issued after Wetterhahn's death, the agency noted the critical need for research laboratories to produce a "protective chemical hygiene plan, which includes adequate guidance on the appropriate selection of personal protective equipment and engineering controls."[5] The memorandum stressed that even "highly placed or very well qualified researchers" do not always possess the most accurate or adequate health and safety information. The memorandum goes on to underscore the need for collaborative relationships between university researchers and health and safety professionals in creating safe and effective laboratory environments.

---

[2] U.S. Department of Labor, Occupational Safety and Health Administration. *Dimethyl Mercury: Hazard Information Bulletin*. Accessed June 30, 2014. http://www.osha.gov/dts/hib/hib_data/hib19980309.html.

[3] Dartmouth Undergraduate Journal of Science. *Remembering Karen Wetterhahn*. May 16, 2008. http://dujs.dartmouth.edu/spring-2008-10th-anniversary-edition/remembering-karen-wetterhahn.

[4] U.S. Department of Labor, Occupational Safety and Health Administration. *Dimethyl Mercury: Hazard Information Bulletin*. Accessed June 30, 2014. http://www.osha.gov/dts/hib/hib_data/hib19980309.html.

[5] Id.

UCLA Incident

Sheharbano (Sheri) Sangji, a staff research assistant at the University of California, Los Angeles (UCLA) working in the lab of Professor Patrick Harran, was attempting to transfer a *tert*-butyllithium solution in hexanes from a reagent bottle to a reaction flask when the plunger of the syringe she was using separated from the barrel, spraying her hands with the pyrophoric compound. Both the *tert*-butyllithium and the hexane ignited, also igniting some additional hexane that had spilled in the commotion and, in the absence of a lab coat, Sangji's highly flammable synthetic sweater caught fire. She initially ran in the opposite direction from the lab safety shower until a co-worker reached her and attempted to extinguish the flames with his lab coat. Another co-worker used water from a nearby sink to finally extinguish the flames. Sangji was rushed to the hospital, but died from her injuries weeks later.

Following Sheri Sangji's death, the State of California's Division of Occupational Safety and Health (Cal/OSHA) undertook an investigation of the accident and the circumstances that led to it. In its report,[6] Cal/OSHA found that Sangji was not following proper safety procedures for handling pyrophoric reagents and had never received adequate training for working with hazardous chemicals required by California code. The report also found that the appropriate personal protective equipment (PPE), specifically lab coats, were not required to be worn. In fact, the report notes that the absence of PPE for researchers was considered "part of the culture"[7] by environmental health and safety (EHS) officials at UCLA.

UCLA took two major steps in response to the Cal/OSHA report. The first was an increase in laboratory safety activities by the EHS office. The EHS office enacted more stringent policies with respect to particularly dangerous chemicals and began inspecting labs more frequently. Laboratory training classes were made mandatory for all laboratory personnel and made available both online and in person on a weekly basis, rather than quarterly as before.[8]

In addition to increasing the role of EHS in laboratory safety, the University of California system created a Center for Laboratory Safety (CLS). The missions of the center, as described on the center's website, are to "sponsor and support research in laboratory safety," "develop and trans-

---

[6] Baudendistel, B. *UCLA Investigation Report*; S 1110-003-09; 2009.

[7] Id., p. 17.

[8] Kemsley, J. N. Learning from UCLA. *Chemical and Engineering News* 2009; 87(31): 29-31, 33-34. http://cen.acs.org/articles/87/i31/Learning-UCLA.html.

fer research into applied best practices," and to "facilitate implementation and optimization of laboratory safety practices."[9]

Nearly two years after Sangji's death, the Los Angeles district attorney's office filed felony criminal charges against both the University of California Regents and Professor Harran for willfully violating occupational health and safety standards. The case against Professor Harran was being heard during the drafting of this report. On June 20, 2014, Harran reached a deferred prosecution agreement with the prosecution, after acknowledging responsibility for the conditions of the laboratory in which the incident occurred. Based on the terms of the agreement, the four criminal counts against Harran will be dropped in five years, if he pays the requested $10,000 fine, fulfills 800 hours of community service at UCLA's hospital, and conducts a summer chemistry course for inner-city high school graduates.[10]

In 2012, the court accepted a plea agreement between the District Attorney and the Regents under which the University of California agreed to strict safety compliance requirements to be enforced by Cal/OSHA. Also, as part of the plea agreement, University of California chemistry departments must compile and maintain standard operating procedures (SOPs) detailing the safety precautions to be taken when using a number of hazardous compounds that are listed in the plea agreement. These SOPs are to be written by senior laboratory staff and then reviewed by "qualified personnel." In addition, the agreement specifies a campus-wide SOP for using pyrophoric materials at UCLA. All SOPs must be made easily available to laboratory personnel, either electronically or in print.

The agreement also prescribes that PPE, including fire-resistant lab coats, must be made available to laboratory researchers. Principal investigators are responsible for reporting any recordable injury[11] or illness to Cal/OSHA and are required to preserve the scenes of any such incidents for subsequent investigation. The University Regents agreed to allow up

---

[9] UC Center for Laboratory Safety. "Overview." Accessed April 1, 2013. http://cls.ucla.edu/.

[10] Whitcomb, D. "UCLA professor ordered to perform community service in fatal lab fire." *Reuters*, June 20, 2014. Accessed June 25, 2014. http://www.reuters.com/article/2014/06/20/us-usa-laboratory-fire-idUSKBN0EV2KW20140620.

[11] According to OSHA, an injury or illness is recordable, if it results in any of the following: death, days away from work, restricted work or transfer to another job, medical treatment beyond first aid, or loss of consciousness. An incident is recordable if it involves a significant injury or illness diagnosed by a physician or other licensed health care professional, even if it does not result in death, days away from work, restricted work or job transfer, medical treatment beyond first aid, or loss of consciousness (Occupational Safety & Health Administration [OSHA]. 2014. Regulations (Standards-29 CFR). Accessed July 1, 2014, https://www.osha.gov/pls/oshaweb/owadisp.show_document?p_table=STANDARDS&p_id=9638).

to three unannounced laboratory inspections by Cal/OSHA per year for 4 years, until 2016.

Texas Tech Incident

Another serious incident, this time involving the shock-sensitive, explosive compound nickel hydrazine perchlorate (NHP), occurred at Texas Tech University in 2010. This incident became the subject of the first investigation of an academic research lab by the U.S. Chemical Safety and Hazard Investigation Board (CSB).[12] The CSB is a nonregulatory government organization that investigates the root cause of chemical accidents, historically focusing on industrial incidents. According to the CSB report,[13] a graduate student attempted to scale up the NHP synthesis, making more than 10 times the amount that had been informally considered an upper limit by his research group. The resulting product, NHP, was clumpy, so the graduate student set out to homogenize the sample by crushing it with a mortar and pestle on an open lab bench. The student removed his safety glasses and subsequently began to crush the NHP "one more time."[14] As the student finished breaking up the clumps, the NHP detonated. The student suffered serious injury to his face, an eye, and his hands, ultimately losing three fingers.

The CSB's analysis in this case was based on the "Swiss-cheese" model of accident causation, where multiple failures align, resulting in an incident. Through this model, the report examined not only the individual mistakes made by the researcher, but the shortcomings across all levels of the organization.

The CSB identified three major flaws in the safety practices at Texas Tech. The first shortcoming, most directly related to the specifics of the accident, was a lack of training and documentation of the physical hazards (e.g., risk of explosion) associated with laboratory research. The second issue identified was a lack of a mechanism for reporting and keeping records of laboratory accidents and "near misses." The CSB argued that without such a mechanism, it is exceedingly difficult to learn from past mistakes. Finally, the CSB found that safety management and oversight were insufficient. The report examined the role of the principal investigator, EHS organization, university leadership, and funding agencies in promoting safety in the laboratory.

---

12 U.S. Chemical Safety and Hazard Investigation Board. *Texas Tech University Laboratory Explosion: Case Study*. Case No. 2010-05-I-TX. Washington, DC, October 19, 2011.

13 Id.

14 Id., p. 15.

Motivation

The incidents at Dartmouth, UCLA, and Texas Tech are notable because of the responses they have garnered, but they by no means represent the totality of reported incidents in U.S. chemistry research labs over the years. In December 2010, a researcher at Northwestern University was injured when an unexpected explosive byproduct was formed in his reaction vessel and detonated.[15] At Yale University in 2011, an undergraduate was killed when her hair was caught in a lathe while she worked alone in a chemistry department machine shop.[16] A less serious accident from the last few years includes the explosion of a glass vial at the University of Colorado at Boulder that caused minor injuries.[17]

Serious accidents in research labs are not limited to academia. In 2008, a researcher at the National Institute of Standards and Technology (NIST) laboratory in Boulder, Colorado, was working with a bottle of radioactive plutonium sulfate tetrahydrate when the bottle broke. The plutonium sulfate got on the researcher's hands and he attempted to wash his hands in the sink before, apparently, realizing the severity of the spill and evacuating. The accident resulted in plutonium being introduced to the Boulder sewer system and the hallway surrounding the lab where the accident happened.[18]

In considering the responsibilities set forth in the statement of task (Box 1-1), understanding the response of oversight organizations to the high-profile accidents at Dartmouth, UCLA, and Texas Tech is critical. Both the Cal/OSHA report on the UCLA incident and the CSB report on the Texas Tech incident point to a deficient safety culture as a primary cause. The three themes from the CSB report are also addressed in the plea agreement between the UC Regents and the State of California. In these cases, the creation of reporting mechanisms, comprehensive SOPs for hazardous compounds, and more comprehensive organizational oversight are emphasized. These incidents have served as new precedents for

[15] Hupp, T., and S. Nguyen. Chemical safety: Synthesis procedure. *Chemical and Engineering News* 2011; 89(2): 2.

[16] Henderson, D., E. Rosenfeld, and D. Serna. Michele Dufault '11 dies in Sterling Chemistry Laboratory accident. *Yale News*, April 13, 2011. Available at http://yaledailynews.com/blog/2011/04/13/michele-dufault-11-dies-in-sterling-chemistry-laboratory-accident/. Accessed September 17, 2012.

[17] University of Colorado Boulder. Glass vial explosion causes evacuation of south wing of CU Engineering Center. News Release, November 30, 2010. Available at http://www.colorado.edu/news/releases/2010/11/30/glass-vial-explosion-causes-evacuation-south-wing-cu-engineering-center. Accessed September 17, 2012.

[18] National Institute of Standards and Technology. *Final Report of the NIST Blue Ribbon Commission on Management and Safety*. U.S. Department of Commerce, Washington, DC, November 2008. Available at http://www.nist.gov/director/blueribbon/upload/final1108.pdf.

the involvement of government agencies and all levels of an organization hierarchy in laboratory safety.

Interest in Safety Culture

The recent serious incidents in academic laboratories have generated significant interest among researchers and safety professionals, demonstrated by frequent editorial articles and blog posts. Numerous editorials in *Chemical and Engineering News* (the news magazine of the American Chemical Society), *Nature, Scientific American,* and other publications have focused on the UCLA accident and the implications of California's response. Blogs maintained by chemists, such as ChemJobber[19] and ChemBark,[20] have also devoted a great deal of effort to discussing both the scientific details of the incidents and ways to improve safety culture to avoid future occurrences.

Some of the discussion of the UCLA and Texas Tech incidents is motivated by the response of regulatory agencies to those accidents. The criminal charges against Professor Harran have sparked intense debate about who bears the ultimate responsibility for laboratory safety. The CSB report on the Texas Tech incident has generated interest not only because it is the first CSB investigation of an academic laboratory or institution, but also because it recommends that funding agencies use safety record as one qualifier for awarding funding. These more controversial topics are rooted in the basic problem of determining how best to promote positive safety culture in academic research labs.

As the details of these incidents continue to be discussed, attention has centered on what could have been done differently in each case. At the same time, a broader discussion of how to prevent serious incidents from occurring in the future and how to give laboratory researchers and emergency personnel the resources to respond appropriately when emergencies do occur is growing.

RECENT WORK

In light of the recent serious safety incidents described above, the American Chemical Society and the National Research Council commissioned or revised reports to emphasize safety in research laboratories.

---

[19] Chemjobber. Available at http://chemjobber.blogspot.com/. Accessed September 17, 2012.

[20] Bracher, P. *Chembark: A Blog About Chemistry & Chemical Research.* Available at http://blog.chembark.com/. Accessed September 17, 2012.

Below is a brief overview of the ACS report on *Creating Safety Cultures in Academic Institutions* and the NRC *Prudent Practices in the Laboratory*.

ACS Report and *Prudent Practices* Discuss Safety Culture in Labs

The American Chemical Society Report

In 2012, ACS assembled a task force to report on *Creating Safety Cultures in Academic Institutions*,[21] which focuses largely on undergraduate teaching laboratories and touches on research labs. It defines safety culture as "a reflection of the actions, attitudes, and behaviors of its members toward safety" and suggests seven characteristics of a strong safety culture: (1) strong leadership and management for safety; (2) continuous learning about safety; (3) strong safety attitudes, awareness, and ethics; (4) learning from incidents; (5) collaborative efforts to build safety culture; (6) promoting and communicating safety; (7) institutional support for funding safety.

With these seven characteristics in mind, the report makes 17 recommendations for academic institutions attempting to improve safety culture. Each recommendation aims to help institutions more strongly demonstrate the seven characteristics of safety culture that the report identifies.

The ACS report focuses on and emphasizes the importance of safety education in undergraduate teaching laboratories. The authors of the report expect that strong safety education during undergraduate studies will translate to graduate students, who form the bulk of the research personnel in academia, with stronger safety ethics and will lead to stronger safety culture in academic labs. In analogy to the responses to the ULCA and Texas Tech incidents, the ACS report emphasizes the need for reporting systems, investigation systems, and a database of safety incidents. The authors suggest that such incident reporting supports continuous learning about safety. In addition to its broader recommendations about strengthening safety culture, the ACS report offers suggestions for the duties that the entire hierarchy of academic laboratories, from university presidents, to principal investigators and faculty, to laboratory staff, might undertake to promote safety.

[21] American Chemical Society Committee on Chemical Safety. *Creating Safety Cultures in Academic Institutions*. American Chemical Society, Washington, DC, 2012: 34.

Prudent Practices in the Laboratory

In 2011, the National Research Council's report, *Prudent Practices in the Laboratory: Handling and Management of Chemical Hazards* (*Prudent Practices*), was updated and included a brief discussion of the role of safety culture in chemical research labs.[22] This report describes safety culture as a "culture of habitual risk assessment, experiment planning, and consideration of worst-case possibilities."[23] *Prudent Practices* notes that researchers leaving academic research labs for industry or government labs are often surprised by the stronger safety culture in industry and government facilities. The report asserts that, "The industrial or government laboratory environment provides strong corporate structure and discipline for maintaining a well-organized safety program where the culture of safety is thoroughly understood, respected, and enforced from the highest level of management down."[24]

In contrast to institutional practices that support a safety culture in industry, academic research laboratories often are embedded in institutions in which safety is rarely discussed outside of targeted training sessions to satisfy regulatory requirements. The turnover in research workers is high; the range of materials and procedures performed by these workers varies considerably across any given institution; and aside from the aforementioned, limited training, many research workers in academic laboratories may have primarily received their safety training from laboratory coursework in chemistry. As a result, safety culture in academic labs faces the difficulty that

> [u]nlike laboratory course work, where training comes primarily from repeating well-established procedures, research often involves making new materials by new methods, which may pose unknown hazards. As a result, workers in academic research laboratories do not always operate from a deep experience base.[25]

This creates challenges for principal investigators and their institutions, particularly in areas of resources and leadership needed to create and sustain safety analyses and practices. As *Prudent Practices* suggests, "[w]hen each principal investigator offers leadership that demonstrates a deep concern for safety, fewer people get hurt."[26] This concern about leadership is a key aspect of safety culture.

---

[22] National Research Council. *Prudent Practices in the Laboratory: Handling and Management of Chemical Hazards, Updated Version*. The National Academies Press, Washington, DC, 2011.

[23] Id., p. 2.

[24] Id., p. 5.

[25] Id., p. 4.

[26] Id.

ORGANIZATION OF THE REPORT

This report is geared to provide guidance to academic research communities on how to strengthen their safety cultures.

Chapter 2 examines safety systems and culture, primarily in the context of sectors outside of academic chemical research. It identifies key themes, principles, and methods that are relevant to laboratory safety and expands on knowledge and experiences in those areas. The chapter culminates by identifying the key attributes of successful safety systems and cultures from the other sectors that are relevant to academic research labs. It cites exemplary approaches and methods utilized in the airline, health care services, and nuclear industries.

Chapter 3 addresses the current state of laboratory safety in chemical research in academic settings. The chapter looks at current practices and attitudes in the context of the current hierarchy of actors involved in laboratory safety, examining current systems that have been utilized and how they work to hinder or raise the safety performance in laboratory research.

Chapter 4 then focuses on understanding laboratory safety dynamics. This final chapter examines the interdependencies that characterize the structure of safety overall, in the context of the current hierarchy of actors involved. After identifying the strengths and weaknesses of the actors, the chapter identifies systems that may be established to raise the overall safety performance of academic research labs.

Chapter 5 presents a series of findings, conclusions, and recommendations that, if followed, can assist institutions in establishing and promoting a culture of safety in academic chemistry research. In keeping with the task at hand, the conclusions and recommendations are focused on chemistry research, but in many cases may be more widely applicable. Chemical hazards can be found in many academic environments, including in the biological sciences, medical schools, many engineering disciplines, and art studios.

2

Safety Systems and Cultures

INTRODUCTION

The goal of this chapter is to summarize approaches and concepts from occupational safety research and practices that are particularly relevant for improving the safety of academic chemistry laboratories. The chapter begins by tracing the development of modern safety practice and the emergence of the concept of safety culture. Next, consideration is given to several different industries that have made good use of modern safety concepts and practices in the face of obvious and significant hazards to people and property. The chapter concludes with a brief discussion of organizational change processes.

The evolution of modern safety management practice is often described in terms of three somewhat overlapping periods or epochs.[1,2] The first phase of development is referred to as the technology period, in which attention was focused on finding and applying engineering or other technological measures to control hazards and prevent work-related injuries.

[1] Hale, A. R., and J. Hovden. Management and culture: The third age of safety. A review of approaches to organizational aspects of safety, health and environment. *Occupational Injury: Risk, Prevention and Intervention*, A. M. Feyer and A. Williamson, eds. Taylor & Francis, London and Bristol, PA, 2003: 129-165.

[2] Hudson, P. Implementing a safety culture in a major multi-national. *Safety Science* 2007; 45(6): 697-722.

The First Epoch: The Technology Period

The hierarchy of hazard controls[3,4] is one of the most enduring products of this period. Within this framework, hazard controls are organized with the highest priority assigned to actions that eliminate the hazard entirely, followed by those that control or otherwise contain the hazard. Lowest priority is assigned to strategies that may be helpful but do not directly remove or alter the hazard, such as warnings or the provision of personal protective equipment (PPE). This basic hierarchy is also reflected in how the Occupational Safety and Health Administration (OSHA) approaches hazard control. OSHA standards typically give preference to engineering controls, followed by administrative controls (training, work rules, etc.), and lastly to the provision of PPE.

As industrial and work systems became more complex, the limits of simple technological solutions quickly became apparent. Many of today's work environments are highly complex, making it difficult to anticipate all possible interactions and possible failures among multiple components and multiple human operators.[5] The traditional view that accidents can be understood in terms of simple linear chains of events has been gradually replaced by a broader systems perspective.[6]

The Second Epoch: The Systems Perspective

The systems perspective represents the second epoch of safety and views accidents and other losses as arising from causal factors that reside at multiple levels within complex sociotechnical systems.[7,8] The concept of human–systems integration (HSI) is central to the systems perspective. HSI focuses on the interaction of people, tasks, and equipment and technology in the pursuit of some goal or set of goals.[9,10] This interaction occurs within, and is influenced by, the broader environmental context.

---

[3] Barnett, R. L., and D. B. Brickman. Safety hierarchy. *Journal of Safety Research* 1986; 17(2): 49-55.

[4] Haddon, W. Energy damage and the ten countermeasure strategies. *Human Factors: The Journal of the Human Factors and Ergonomics Society* 1973; 15(4): 355-366.

[5] Leveson, N. A new accident model for engineering safer systems. *Safety Science* 2004; 42(4): 237-270.

[6] Perrow, C. *Normal Accidents: Living with High Risk Technologies (Updated).* Princeton University Press, Princeton, NJ, 2011.

[7] Rasmussen, J. Risk management in a dynamic society: A modelling problem. *Safety Science* 1997; 27(2): 183-213.

[8] Reason, J. Human error: Models and management. *BMJ* 2000; 320(7237): 768-770.

[9] Booher, H. R. *Handbook of Human Systems Integration,* Vol. 23. John Wiley & Sons, Hoboken, NJ, 2003.

[10] Czaja, S. J., and S. N. Nair. Human factors engineering and systems design. *Handbook of Human Factors and Ergonomics,* 3rd Ed., John Wiley & Sons, Hoboken, NJ, 2006: 32-49.

HSI acknowledges that people differ in terms of their cognitive, perceptual, and physical capabilities, and that these capabilities influence how they interact with different tasks, and equipment and technology. These interactions take place in some larger environment or set of environments, which also have their own characteristics that are capable of either facilitating or impeding the successful use of equipment and/or technology and completion of tasks.

Work systems are basically open systems; that is, they can be influenced by both internal and external factors. For example, external factors such as economic conditions or competitive pressures have the potential to impact safety, either positively or negatively. Also inherent in the systems perspective is the idea that some systems may require defenses in depth or redundant controls at different points or levels within the system. The distinction between active and latent failures is also pertinent to the systems approach.[11] Errors or mistakes made by frontline workers or researchers are frequently referred to as active failures, and these active failures are often the result of actions or decisions taking place (or not taking place) at higher levels of the organization (latent failures). Effective and permanent solutions to safety problems at the lab bench level often require the identification and elimination of these latent failures, which sometimes are hidden or lie dormant within organizations for long periods of time before contributing to adverse events.

An important feature of the systems approach is the acknowledgment that entire systems can degrade subtly or drift toward failure. Various specialized analytic tools or techniques have been developed to help identify and guard against such shifts in system integrity. These specialized tools include a variety of different risk assessment methodologies such as fault tree analysis and failure modes and effects analysis. Most of these techniques can be used to analyze system vulnerabilities or to reconstruct and understand why failures occurred. Systems safety also makes use of safety audits and other techniques that can be used to monitor system performance and provide early detection of changes in key system parameters.

The Third Epoch: Safety Culture

The emphasis on culture, specifically safety culture, represents the third epoch of modern safety management. This shift or expansion came about from the realization that it is not enough to provide safe equipment, systems, and procedures if the culture of the organization does not

[11] Reason, J. Achieving a safe culture: Theory and practice. *Work & Stress* 1998; 12(3): 293-306.

encourage and support safe working. Hudson[12] argues that safety culture is probably the most important issue in modern thinking and practice in safety. The investigative report[13] that followed the Chernobyl nuclear disaster is usually credited with introducing the concept of safety culture. Since that time, safety culture has been a prominent feature in the investigation and analysis of most major or catastrophic accidents, including, for example, the recent Deepwater Horizon oil spill. In essence, safety culture forms the organizational context in which all actions pertinent to safety occur.

Although there is no uniform definition offered in the literature, "safety culture" arose from a more general understanding of organizational culture. Edgar Schein, a psychologist credited with pioneering the field of organizational culture, explains that culture embodies values, beliefs, and underlying assumptions.[14] Schein describes culture as something that is developed over time by a group as it "solves its problems of external adaptation and internal integration, which has worked well enough to be considered valid, and therefore to be taught to new members as the correct way to perceive, think, and feel in relation to those problems."[15] Table 2-1 presents a summary of three models of organizational culture generally accepted by behavioral and social scientists. Taken further, safety culture is most often identified by an organization's response to or prevention of workplace accidents.

Safety culture, as typically defined, refers to the organization's shared values, assumptions, and beliefs specific to workplace safety, or more simply, the relative importance of safety within the organization.

Numerous attempts have been made to identify the key attributes or characteristics of a positive safety culture, and although the various frameworks differ in the details, there are clearly more similarities than differences.[16,17,18] For example, virtually all discussions of safety culture highlight the fundamental importance of management commitment and active involvement. Frameworks also emphasize the importance of com-

---

[12] Hudson, P. Implementing a safety culture in a major multi-national. *Safety Science* 2007; 45(6): 697-722.

[13] International Atomic Energy Agency. *Safety Culture: A Report by the International Nuclear Safety Advisory Group.* Safety Series No. 75-INSAG-4. Vienna, Austria, 1991.

[14] Schein, E. H. *Organizational Culture and Leadership.* John Wiley & Sons, Hoboken, NJ, 2010.

[15] Id., p. 18.

[16] DeJoy, D. M. Behavior change versus culture change: Divergent approaches to managing workplace safety. *Safety Science* 2005; 43(2): 105-129.

[17] Hopkins, A. Studying organisational cultures and their effects on safety. *Safety Science* 2006; 44(10): 875-889.

[18] Wiegmann, D. A., H. Zhang, T. L. von Thaden, G. Sharma, and A. Mitchell Gibbons. Safety culture: An integrative review. *International Journal of Aviation Psychology* 14(2): 117-134.

TABLE 2-1 Three Models of Organizational Culture

	Level		
Originator	Most Accessible	Intermediate	Deepest
Schein (1985)	Behaviors and artefacts	Beliefs and values	Underlying assumptions
Rousseau (1988, 1990)	Observable artefacts (e.g., company logo); observable patterns of behavior	Behavioral norms, which can be inferred from observed behaviors; values, as expressed consciously by organization members	Fundamental assumptions—core values that may not be articulated
Deal and Kennedy (1986); Lundberg (1990)	Manifest level—symbolic artefacts, language, stories, rituals, normative behaviors	Strategic level—strategic beliefs	Core level—ideologies, values, assumptions

Adapted from Glendon, A. I., and Stanton, N. A. Perspectives on safety culture. *Safety Science* 2000; 34(1): 193-214.

munication and the free exchange of safety-related information, especially the freedom of all members to report hazards and to be heard on matters involving safety. Positive safety cultures also place high importance on hazard identification and control as well as continuous learning and improvement. To a considerable extent, achieving a safety culture that emphasizes learning and improvement requires a culture that seeks and values information and that assigns greater importance to problem solving than blame assignment. Obviously, a positive safety culture is one in which a high relative importance is assigned to safety all the time, not just when it is convenient or does not threaten personal or institutional productivity goals. However, the strongest, most positive safety culture is established when all members at all levels of the organization basically agree on the importance of safety. However, particularly within large or loosely structured organizations, there are many opportunities for "disconnects" to occur as to the primacy of the safety mission. Such variability or heterogeneity can easily undermine safety performance. Disconnects also can occur in work situations where individual members or workgroups have relatively high levels of discretion in how their work is planned and executed.[19,20,21,22,23]

Besides identifying the core traits of positive safety cultures, other researchers have sought to create taxonomies of safety culture types. These taxonomies can be used by organizations for purposes of self-assessment and change or they can be used to help verify and refine the key attributes of safety culture. Westrum developed a taxonomy consisting of three types of cultures that were distinguished primarily in terms of how information is handled.[24] His three culture types were pathological, bureaucratic, and generative. Pathological cultures are basically power-oriented and information is viewed as a personal resource to be guarded. Bureaucratic cultures are heavily rule-oriented, and information is often not welcome or is ignored. Generative cultures, on the other hand, are more performance-oriented. In such cultures, information is welcomed, and efforts are made to get the right information to the right person at the right time. The pathological culture is a blame-type culture, the bureaucratic culture is a compliance-type culture, and a generative culture is a more proactive and positive culture.

Others have extended this basic typology. Parker and colleagues, in particular, describe five culture types: pathological, reactive, calculative, proactive, and generative. They summarize the five cultures as follows: pathological, "who cares as long as we are not caught"; reactive, "safety is important: we do a lot every time we have an accident"; calculative, "we have systems in place to manage all hazards"; proactive, "we try to anticipate safety problems before they arise"; and generative, "health, safety, and environment is how we do business around here."[25] These authors also outline how each culture type would likely handle various

---

[19] Hage, J. and M. Aiken. Routine technology, social structure, and organizational goals. *Administrative Science Quarterly* 1969; 14: 366-378.

[20] Zohar, D. Safety climate: Conceptual and measurement issues. In J. C. Quick and L. E. Tetrick (eds). *Occupational Health Psychology,* 2nd ed. American Psychological Association, Washington, DC, 2011: 141-164.

[21] Zohar, D. Modifying supervisory practices to improve sub-unit safety: A leadership-based intervention model. 2002; *Journal of Applied Psychology* 87: 156-163.

[22] Zohar, D., The effects of leadership dimensions, safety climate, and assigned priorities on minor injuries in work groups. 2002; *Journal of Organizational Behavior* 23: 75-92.

[23] Kines, P., L. P. Andersen, J. Dyreborg, and D. Zohar. Improving construction site safety through leader-based verbal safety communication. 2010; *Journal of Safety Research* 41: 399-406.

[24] Westrum, R. A typology of organisational cultures. *Quality and Safety in Health Care* 2004; 13(S2): ii22-ii27.

[25] Parker, D., M. Lawrie, and P. A. Hudson. A framework for understanding the development of organisational safety culture. *Safety Science* 2006; 44(6): 551-562.

aspects of safety management, such as safety audits and reviews, work planning, and handling contractors.

Mindfulness and Situational Awareness

Mindfulness is a psychological quality that involves bringing one's complete attention to the present experience on a moment-to-moment basis in a nonjudgmental way. The mindless following of routine and other automatic behaviors leads to error, pain, and a predetermined course of life. To be mindful stresses process over outcomes, allowing free rein for intuition and creativity, and opens us to new information and perspectives. When applied to safety, the concept of mindfulness extends to groups as well as individuals. Indeed, collective mindfulness is an important factor in achieving high levels of safety in high-hazard situations.[26]

The development of situational awareness requires mindfulness. While there are many definitions of situational awareness, Endsley's is probably the most commonly used: "the perception of elements in the environment within a volume of time and space, the comprehension of their meaning, and the projection of their status in the near future."[27] Situational awareness is commonly used in complex domains, such as air traffic control or surgery. It is often called upon in time-critical situations in which choices have to be made quickly by decision makers, with the support of other team members and a myriad of information coming from other sources. Situational awareness relates more to achieving immediate tactical objectives than to long-term objectives.

The development of sense-making requires situational awareness. Sense-making addresses more long-term strategic issues than situational awareness. Klein and colleagues define sense-making as "a motivated, continuous effort to understand connections (which can be among people, places and events) in order to anticipate their trajectories and act effectively."[28] It is a constant process of acquisition, reflection, and action.

Their view of the process is one shared of many organizational theorists (e.g., Westrum[29]) where, in a large organization, various people may hold different pieces of data, and different levels of awareness of events, that are all critical to the success of a given project. Sense-making is deeply

---

26 Langer, E. J. *Mindfulness*. Addison Wesley Longman, Boston, MA, 1989.

27 Endsley, M. R. Measurement of situation awareness in dynamic systems. *Human Factors: The Journal of the Human Factors and Ergonomics Society* 1995; 37(1): 65-84.

28 Klein, G., B. M. Moon, and R. R. Hoffman. Making sense of sensemaking, 1: Alternative perspectives. *IEEE Intelligent Systems* 2006; 21(4): 70-73.

29 Westrum, R. A typology of organisational cultures. *Quality and Safety in Health Care* 2004; 13(S2): ii22-ii27.

related to a process of "socialization," whereby those with ideas and data share them with others in an effort to actively disseminate information and build consensus. Klein and colleagues'[30] view of sense-making is a process that is both personal and shared, one that takes place over a long period of time, and one that is heavily dependent on a perspective or point of view.[31]

INVOLVEMENT, GROUPS, AND TEAMS

Promoting workers' involvement at all levels can be an effective way to help build and sustain a positive safety culture.[32] It is especially important for improving the exchange of safety-related information, fostering collective mindfulness and sense-making, empowering workers to speak up and share what they know, and creating a learning and improvement focus. Involvement also can facilitate the successful implementation of new programs and initiatives.[33] In many work situations, managers or other leaders are often unaware of frontline safety problems. Getting frontline workers involved by thinking and talking about safety is one way to address this problem and leverage the expertise that these workers possess. As Susan Silbey argues, "Lower-level actors are often repositories of critical information, yet are often unable to persuade higher-ups in the organization of either the credibility of their knowledge or the relevance of their perspective."[34] Worker involvement has been linked to better safety outcomes in a number of work settings, including chemical plants,[35] oil and gas extraction,[36] and health care.[37] Health and safety committees are perhaps the most frequently employed worker involvement

[30] Klein, G., B. M. Moon, and R. R. Hoffman. Making sense of sensemaking, 1: Alternative perspectives. *IEEE Intelligent Systems* 2006; 21(4): 70-73.

[31] Kolko, J. *Sensemaking and Framing: A Theoretical Reflection on Perspective in Design Synthesis.* frog design & Austin Center for Design, Austin, TX, 2010. Available at http://www.designresearchsociety.org/docs-procs/DRS2010/PDF/067.pdf.

[32] Simard, M., and A. Marchand. Workgroups' propensity to comply with safety rules: The influence of micro-macro organisational factors. *Ergonomics* 1997; 40(2): 172-188.

[33] Lawler, E. J. Affective attachments to nested groups: A choice-process theory. *American Sociological Review* 1992; 57(3): 327-339.

[34] Silbey, S. S. Taming Prometheus: Talk about safety and culture. *Annual Review of Sociology* 2009; 35 (2009): 341-369.

[35] Hofmann, D. A., and A. Stetzer. A cross-level investigation of factors influencing unsafe behaviors and accidents. *Personnel Psychology* 1996; 49(2): 307-339.

[36] Mearns, K., S. M. Whitaker, and R. Flin. Safety climate, safety management practice and safety performance in offshore environments. *Safety Science* 2003; 41(8): 641-680.

[37] Singer, S., S. Lin, A. Falwell, D. Gaba, and L. Baker. Relationship of safety climate and safety performance in hospitals." *Health Services Research* 2009; 44(2 Pt 1): 399-421.

strategy specific to safety,[38] but employee involvement can take many different forms.

Behavioral and organization scientists have devoted considerable attention to various types of worker involvement approaches. High-performance work systems and high-involvement work processes (HIWPs) are two approaches that have received considerable research attention. Both approaches involve sets of work practices designed to leverage employee motivation and creativity, and in some sense represent reactions against scientific management and its centralization of decision making and problem solving at the management level.[39] Using HIWPs as an example, Edward Lawler proposed a framework consisting of four HIWPs: power (P), information (I), reward (R), and knowledge (K).[40] These four processes (PIRK) are intended to be mutually reinforcing. HIWPs empower workers to make more decisions on the job, provide them with the information and knowledge they need for decision making, and reward them for doing so.

Employee empowerment is a central feature of most high-performance and high-involvement models and frameworks. Social support within the workgroup and from managers and supervisors is important in empowering employees and giving them "voice" in safety matters. This permits them to speak out and to modify or halt work that they consider too risky.[41,42] Empowerment is also a key attribute of high-reliability organizations (HROs). HROs strive for constant safety mindfulness, and there is deference to expertise whereby authority migrates down the command structure to whomever has the most pertinent knowledge or the best perspective for understanding and solving a problem.[43,44] This priority on safety mindfulness often extends to recognizing and rewarding people

---

[38] Dunlop Commission. *Report and Recommendations of the Commission on the Future of Worker-Management Relations.* U.S. Department of Labor and Department of Commerce, Washington, DC, 1994: Section II.

[39] Boxall, P., and K. Macky. Research and theory on high-performance work systems: Progressing the high-involvement stream. *Human Resource Management Journal* 2009; 19(1): 3-23.

[40] Lawler, E. E. *High-Involvement Management*. Jossey-Bass, San Francisco, CA, 1986.

[41] Conchie, S. M., P. J. Taylor, and I. J. Donald. Promoting safety voice with safety-specific transformational leadership: The mediating role of two dimensions of trust. *Journal of Occupational Health Psychology* 2012; 17(1): 105-115.

[42] Tucker, S., N. Chmiel, N. Turner, M. S. Hershcovis, and C. B. Stride. Perceived organizational support for safety and employee safety voice: The mediating role of coworker support for safety. *Journal of Occupational Health Psychology* 2008; 13(4): 319-330.

[43] Roberts, K. H. Some characteristics of one type of high reliability organization. *Organization Science* 1990; 1(2): 160-176.

[44] Rochlin, G. I. Safe operation as a social construct. *Ergonomics* 1999; 42(11): 1549-1560.

even when their safety-related concerns prove to be inaccurate or not well founded.

Forming workgroups or teams is another strategy for increasing employee involvement and empowerment. Effective teams have shared mental models and group situation awareness; they also efficiently process, share, and use information. These attributes are especially important in emergency and high-stress situations where performance must be adapted to cope with rapidly changing or unexpected conditions. Simulations and other interactive training activities allow team members to operate as a team while training, engage in the social, cognitive, and behavioral processes of team performance, and receive feedback based on their performance. Training approaches such as Cockpit Resource Management (CRM) in aviation[45] directly foster team skills, including assertiveness, maintaining shared situation awareness, and communication. Considerable research, much involving cockpit crews, but also some in health care, underscores the importance of group processes and group cohesion in overall safety.[46,47]

KNOWLEDGE FROM OTHER SAFETY SYSTEMS

This section describes and discusses several examples of industries that have adopted many of the principles and approaches described above with the primary goal of improving safety performance. Many of these same industries are those utilizing complex technologies, obvious inherent hazards, and the potential for experiencing serious or even catastrophic losses. The experiences of these industries may be relevant to experimental research with chemicals.

Aviation

Aviation safety was given a boost by a National Aeronautics and Space Administration (NASA)-sponsored workshop, "Resource Management on the Flightdeck," in 1979. This conference was the outgrowth of NASA's research into the causes of commercial air transport accidents. The research presented at this meeting identified the human error aspects of the majority of air crashes as failures of communication, decision making, and leadership. At this meeting, the label Cockpit Resource Manage-

---

[45] Wiener, E. L., B. G. Kanki, and R. L. Helmreich, eds. *Cockpit Resource Management*. Gulf Professional Publishing, Houston, TX, 1993.

[46] Clarke, S. The contemporary workforce: Implications for organisational safety culture. *Personnel Review* 2003; 32(1): 40-57.

[47] Helmreich, R. L., and A. C. Merritt. *Culture at Work in Aviation and Medicine: National, Organizational and Professional Influences*. Ashgate Publishing, Surry, UK, 2001.

ment (CRM) was applied to the process of training crews to reduce pilot error by making better use of the human resources on the flightdeck.[48]

The first comprehensive cockpit or CRM program was initiated by United Airlines (UAL) in 1981 following a devastating UAL accident in Portland, Oregon. As CRM developed, training emphasis was placed increasingly on group dynamics. The new courses dealt with more specific aviation concepts related to flight operations, became more modular, and became more team-oriented in nature. Basic training conducted in intensive seminars included concepts such as team building, briefing strategies, situational awareness, and stress management. Specific modules addressed decision-making strategies and breaking the chain of errors that can result in catastrophe. Much of CRM addresses communication processes and power and knowledge differentials within interdependent work groups.

Reporting

Fortunately, actual incidents involving injury and damage to property are relatively rare, even in very high hazard environments. Much more common, however, are near misses. Near misses are incidents or events that could have resulted in injuries or other adverse consequences, but fortunately did not.[49,50] The loss potential of a near miss is quite real; the difference between a near miss and an actual accident often amounts to a fraction of a second or a fraction of an inch. With a near miss, some combination of unsafe conditions and/or behaviors existed and a sequence of events unfolded that could have led to adverse outcomes. Although often ignored, near misses represent an important data source for learning and prevention. Near misses are often symptomatic of some type of system vulnerability or degradation, which, if uncorrected, may cause serious problems in the future. They might best be viewed as instructive. The importance and usefulness of reporting and tracking near misses has gained broad recognition in many areas of safety practice. Near miss reporting can be a useful part of the surveillance and monitoring component of a comprehensive safety management system.

---

[48] Helmreich, R. L., A. C. Merritt, and J. A. Wilhelm. The evolution of Crew Resource Management training in commercial aviation. *International Journal of Aviation Psychology* 1999; 9(1): 19-32.

[49] Jones, S., C. Kirchsteiger, and W. Bjerke. The importance of near miss reporting to further improve safety performance. *Journal of Loss Prevention in the Process Industries* 1999; 12(1): 59-67.

[50] Wright, L., and T. Van der Schaaf. Accident versus near miss causation: A critical review of the literature, an empirical test in the UK railway domain, and their implications for other sectors. *Journal of Hazardous Materials* 2004; 111(1): 105-110.

The Federal Aviation Administration (FAA) has perhaps the best-known near-miss reporting system in the United States. This system, the Aviation Safety Reporting System (ASRS), allows pilots and other personnel to confidentially report near misses and other close calls. The reporting system was first established in 1976. The ASRS is confidential and independent, and near-miss reports cannot normally be used in any FAA enforcement actions. Independence is achieved by having the system maintained by NASA. Those making reports do not have to (but may) provide their name and contact information. Once staff analysts are satisfied with the information contained in a report, contact information is removed from the report. ASRS analysts may identify hazardous situations from reports and issue "Alert Messages" to organizations within the aviation sector. The database of reports is also used for research (modeling, trending, root-cause taxonomies, etc.) and other purposes intended to better inform the aviation community and benefit safety. ASRS reports are available from NASA's ASRS website. The database is searchable and available to the public.

Near-miss reports can be submitted electronically or by mail. The report form includes space for describing the event or situation. Cues are provided on the form encouraging reporters to address causal and contributing factors, the sequence of events involved, and any human performance factors involved. Much of the remainder of the form consists of sets of checkboxes that collect information specific to different contributing factors and conditions. Since its inception, over 1 million reports have been submitted; including over 70,000 reports in 2012. The system is promoted as confidential, voluntary, and nonpunitive. The National Firefighter Near-Miss Reporting System is of more recent origin. This system is based closely on the ASRS and is supported and funded by the U.S. Department of Homeland Security as part of the Assistance to Firefighters Grants Program. Near-miss reporting systems have been advocated for a number of other industries, including the chemical process industry[51] and the health care industry.[52]

A commitment by the entire scientific community to promote an effective near-miss reporting system might ultimately be productive in practice. The difficulty lies in the necessary level of detail of the reported chemicals and materials used in the potential hazard as well as the documentation of the level of experience of those involved. However, the

[51] Phimister, J. R., U. Oktem, P. R. Kleindorfer, and H. Kunreuther. Near-miss incident management in the chemical process industry. *Risk Analysis* 2003; 23(3): 445-459.

[52] Institute of Medicine. *Patient Safety: Achieving a New Standard of Care*. The National Academies Press, Washington, DC, 2004.

chemistry community, with all its levels of expertise, has a great opportunity to optimize an anonymous near-miss reporting system.

Health Care

Since the Institute of Medicine's (IOM's) 2000 publication of *To Err Is Human*,[53] the health care community has given a great deal of attention to patient safety. *To Err Is Human* and its 2001 follow-up publication, *Crossing the Quality Chasm*,[54] both concluded that health care is not as safe as it should be and suggested that between 44,000 and 98,000 patients are killed in hospitals in the United States every year—medical errors that could have been prevented. The highest error rates with serious consequences are most likely to occur in intensive care units, operating rooms, and emergency departments.

The authors hypothesize that error rates are so high because of the decentralized and fragmented nature of health care in the United States. They propose that medical errors are not the result of individual recklessness, but result from faulty systems, and the pressures that lead people to make mistakes or not prevent them from happening.

Crossing the Quality Chasm recommends redesigning the American health care system and provides specific direction for policy makers, health care leaders, clinicians, regulators, purchasers, and others. Health care providers are asked to adopt a shared vision of six specific aims for improvement. These aims are built around the core need for health care to be

- Safe: avoiding injuries to patients from the care that is intended to help them.
- Effective: providing services based on scientific knowledge to all who could benefit, and refraining from providing services to those not likely to benefit.
- Timely: reducing waits and sometimes harmful delays for both those who receive and those who give care.
- Efficient: avoiding waste, including waste of equipment, supplies, ideas, and energy.
- Equitable: providing care that does not vary in quality because

[53] Institute of Medicine. *To Err Is Human: Building A Safer Health System*. National Academy Press, Washington, DC, 2000.

[54] Institute of Medicine. *Crossing the Quality Chasm: A New Health System for the 21st Century*. National Academy Press, Washington, DC, 2001.

of personal characteristics such as gender, ethnicity, geographic location, and socioeconomic status.[55]

In response to the IOM and Congress, in 2001 the Agency for Healthcare Research and Quality (AHRQ) renamed its Center for Quality Measurement and Improvement, the Center for Quality Improvement and Patient Safety. This was step 1 in AHRQ's efforts to refocus and concentrate in one unit its research and implementation activities devoted to safety in health care.

In 2008, AHRQ published *Becoming a High Reliability Organization: Operational Advice for Hospital Leaders.* In it, they spelled out Sutcliffe and Weick's[56] high-reliability characteristics:

- **Sensitivity to operations**. HROs recognize that manuals and policies constantly change and are mindful of the complexity of the systems in which they work. HROs work quickly to identify anomalies and problems in their system to eliminate potential errors. Maintaining situational awareness is important for staff at all levels because it is the only way anomalies, potential errors, and actual errors can be quickly identified and addressed.
- **Reluctance to simplify.** HROs refuse to simplify or ignore the explanations for difficulties and problems that they face. These organizations accept that their work is complex and do not accept simplistic solutions for challenges confronting complex and adaptive systems. They understand that their systems can fail in unexpected ways that have never happened before and that they cannot identify all the ways in which their systems could fail in the future.
- **Preoccupation with predicting potential failures.** HROs are focused on predicting and preventing catastrophes rather than reacting to them. These organizations constantly entertain the thought that they may have missed something that places patients at risk. Near misses are viewed as opportunities to improve current systems by examining strengths, determining weaknesses, and devoting resources to improve and address them.
- **Deference to expertise.** HROs cultivate a culture in which team members and organizational leaders defer to the person with the most knowledge relevant to the issue they are confronting. The most generally experienced person or the person highest in the

[55] Id.

[56] Sutcliffe, K. E., and K. M. Weick. *Managing the Unexpected: Assuring High Performance in an Age of Complexity,* Wiley India, New Delhi, 2006.

organizational hierarchy does not necessarily have the information most critical to responding to a crisis.
- **Resilience.** HROs pay close attention to their ability to quickly respond to and contain errors and recover when difficulties occur. Thus, systems can function despite setbacks.[57]

The health care industry engages in other activities designed to promote patient safety, such as the National Patient Safety Foundation and the Lucean Leape Institute. Have these activities improved patient safety? It is hard to know because, as is true for aviation, the obvious indicators, such as morbidity and mortality, are influenced by many variables. Unlike aviation, however, there are enough incidents to obtain reliable statistical metrics.

Stelfox and co-workers searched MEDLINE for articles on patient safety and medical error from November 1, 1994, to November 1, 2004, and examined federal funding of patient safety research from 1995 to 2004.[58] The rate of publication of patient safety research was significantly ($p < .01$) higher after publication of *To Err Is Human* than before. Prior to the book's publication, patient safety publications were overwhelmingly about malpractice; however, after publication, they were overwhelmingly about culture. Research support was also higher after the book's publication. Many of the activities described are relevant to the academic chemistry community and are worth consideration.

Industrial Research Facilities

Foundations in Regulation Mandatory Safety and Health Standards

Because of their size and scope of work, industrial research facilities are most often subject to OSHA regulations. As such, their fundamental laboratory safety principles are driven by 29 CFR § 1910.1450, *Occupational Exposure to Hazardous Chemicals in Laboratories standard (the "Laboratory standard")*, and various other hazard-specific standards, such as 29 CFR § 1910.1030, *Bloodborne pathogens*, and § 1910.101, *Compressed gases*. These

---

[57] Hines, S., K. Luna, J. Lofthus, M. Marquardt, and D. Stelmokas. *Becoming a High Reliability Organization: Operational Advice for Hospital Leaders*. AHRQ Publication No. 08-0022. Agency for Healthcare Research and Quality, Rockville, MD, April 2008. Available at http://www.ahrq.gov/professionals/quality-patient-safety/quality-resources/tools/hroadvice/hroadvice.pdf.

[58] Stelfox, H. T., S. Palmisani, C. Scurlock, E. J. Orav, and D. W. Bates. The "To Err Is Human" report and the patient safety literature. *Quality and Safety in Health Care* 2006; 15(3): 174-178.

standards target individual hazards and follow a basic logic: identify the hazard, evaluate the hazard, train workers, and control the hazard.

Industrial research facilities are often associated with production facilities. These facilities are also required to follow OSHA's § 1910.119, *Process safety management of highly hazardous chemicals (aka Process Safety Standard)*, if they utilize processes that involve specific chemicals above the threshold quantities listed in the standard. While the purpose of the standard is to prevent or minimize the consequences of catastrophic chemical releases, the standard's requirements have the added benefit of increasing individual safety. Because bench-scale research facilities are used to test production ideas prior to scale-up, these research facilities often utilize modified versions of the Process Safety Standard requirements. Examples of the standard requirements that can be modified for research activities include the following:

- Hazard analyses are conducted at the process level utilizing methods similar to those prescribed by the standard (i.e., what-if, checklist, hazard and operability study, failure modes and effects analysis, or fault tree analysis).
- The analysis addresses hazards of the process, the identification of previous incidents that could lead to catastrophic consequences, the identification of engineering and administrative controls, the consequences of failure, and human factors.
- Employees are involved in the hazard analysis.
- Systems are designed to comply with code requirements and generally accepted good engineering practices.
- Written operating procedures are utilized to provide clear instructions for safely conducting an activity.
- Employees are trained in the safe conduct of the process as well as the emergency procedures required should a failure occur.
- The hazard analysis is periodically revisited and modified if changes are anticipated in the process.
- Inspections and testing are used to identify drift from expected performance.

Research institutions that incorporate these principles into their research and development activities move from using a predefined set of controls and standard laboratory practices to incorporating a systematic approach to recognition, evaluation, and control of high-hazard activities into their way of doing business.

Growth through Adoption of Consensus Safety and Health Standards

In 1989, OSHA announced its intent to publish voluntary guidelines that employers could use to develop safety and health management programs. The guidelines were not well received by the public, and OSHA ultimately withdrew its intent. Since then, consensus standards have been developed that incorporate many of the principles laid out by OSHA in its proposed rulemaking. Most notably, the American Industrial Hygiene Association served as secretariat in cooperation with the American Society of Safety Engineers to publish ANSI Z10, *American National Standard for Occupational Health and Safety Management Systems,* and a number of cooperating national standards bodies from around the world assisted in the development of OHSAS 18001, *Occupational Health and Safety.*[59] Many industrial research facilities have voluntarily adopted these standards because (1) they find value in applying the management system approach to improve organizational performance, (2) the basic structure of the consensus standards creates a framework with which the institution can demonstrate compliance with a number of specification standards, and (3) they see implementation as a competitive advantage in the international marketplace.

The standards incorporate principles engineered to integrate health and safety into the fabric of an organization rather than to exist as a standalone set of processes or standards. Marked differences between these standards and more traditional regulations include

- Management leadership and commitment;
- Clearly defined roles, responsibilities, accountabilities, and authorities;
- Identification of institutional risks, followed by performance objective and resource allocations;
- Incident investigation;
- Focus on preventive actions;
- Clear involvement by management in the review of system performance; and
- Voluntary assessment by external registration bodies.

Institutions that voluntarily follow these standards are consciously or subconsciously agreeing to modify their culture.

[59] BS OHSAS 18001 Occupational Health and Safety Management. http://www.bsigroup.com/en-GB/ohsas-18001-occupational-health-and-safety/. Accessed July 28, 2014.

Nuclear Industry

Safety has been a primary consideration in the nuclear industry from the very start, beginning with the Manhattan Project during World War II. Part of this concern was obviously related to the magnitude of the hazards involved and the potential for serious or catastrophic harm, not just to workers, but to the general public and the environment. The multidisciplinary nature of the enterprise also contributed to a heightened focus on safety and high regulation of the industry. Harnessing nuclear power and building reactors was very much a multidisciplinary enterprise, requiring that scientists and engineers from multiple disciplines work together to meld their different perspectives on design and construction. Despite these precautions,[60] events such as Three Mile Island and Chernobyl have served to reinforce these concerns. The nuclear industry from the very beginning has been a highly regulated industry and the safety of nuclear energy remains a visible and sometimes volatile public policy issue.

Perrow emphasizes that some technological systems possess certain characteristics that make them inherently hazardous.[61] From his perspective, two dimensions are particularly important: complexity and tight coupling. Complex systems, defined as those involving multiple interactions and many different components, are inherently more susceptible to unanticipated outcomes and mistakes than operations involving simple linear interactions. Tight coupling exists when there is little opportunity to correct or counteract errors or malfunctions once they occur. In a tightly coupled system, minor errors or failures can rapidly cascade out of control and produce serious consequences before corrective measures can be taken. Nuclear power plants are both very complex and tightly coupled. The typically simple task of keeping track of system status can be a challenge in such systems. Indeed, this very problem was an important contributing factor in the Three Mile Island incident.

Initial approaches to controlling risk in the nuclear industry primarily focused on providing defense in depth, redundancies, and wide safety margins. These actions were soon supplemented by the application of quality assurance techniques in design and manufacture and the use of continuous testing, inspection, and maintenance to keep system performance within design limits. As the industry continued to develop, systems safety techniques such as fault tree analysis and event trees were utilized to estimate risk and identify system weakness and vulnerabili-

[60] Keller, W., and M. Modarres. A historical overview of probabilistic risk assessment development and its use in the nuclear power industry: A tribute to the late Professor Norman Carl Rasmussen. *Reliability Engineering & System Safety* 2005; 89(3): 271-285.

[61] Perrow, C. *Normal Accidents: Living with High Risk Technologies (Updated)*. Princeton University Press, Princeton, NJ, 2011.

ties. Probabilistic risk assessment has become an important component of safety management within this industry. The Three Mile Island incident provided an important stimulus for increased attention to general issues such as operator training and human factors more generally. It also led to increased application of accident scenarios, simulation techniques, and the monitoring and investigation of near misses and other precursor events. Greater acceptance was given to the idea that even minor events can cause major losses. As the industry has matured, there has been increased acknowledgment that each power plant is unique and may have its own specific vulnerabilities. The nuclear industry, along with other high-hazard industries, has also come to realize the importance of "upstream" organizational and managerial factors in accident causation and safety performance.[62,63]

The safety culture concept originated in the nuclear industry in the aftermath of the Chernobyl disaster in 1986.[64] As discussed previously, in 1991, the International Atomic Energy Agency (IAEA) issued a comprehensive report on safety culture, defining it for the nuclear industry. Safety culture was defined as "that assembly of characteristics and attitudes in organizations and individuals which establishes that, as an overriding priority, nuclear plant safety issues receive the attention warranted by their significance."[65] The definition was crafted to emphasize both organizational and individual commitment, management responsibility for policy, and the operational framework and staff responsibility for commitment and competence. The IAEA also offered quite detailed guidance for establishing and managing a positive safety culture. The U.S. Nuclear Regulatory Commission also has issued several reports and statements pertinent to safety culture. Some of the earlier documents focused on assigning top priority to safety and making sure that employees can raise safety concerns without fear of retaliation. In 1998, the Nuclear Regulatory Commission initiated its Reactor Oversight Process.[66] This report included three cross-cutting themes that were intended to apply to all aspects of safety: human performance, management attention to safety

---

[62] Flin, R., K. Mearns, P. O'Connor, and R. Bryden. Measuring safety climate: Identifying the common features. *Safety Science* 2000; 34(1): 177-192.

[63] Weick, K. E., K. M. Sutcliffe, and D. Obstfeld. Organizing for high reliability: Processes of collective mindfulness. *Crisis Management*, Vol. 3, A. Boin, ed. Sage, London, UK, 2008: 81-123.

[64] Nuclear Energy Agency. *Chernobyl and the Safety of Nuclear Reactors in OECD Countries: Report*. Organisation for Economic Co-operation and Development, 1987.

[65] International Nuclear Safety Advisory Group. *Management of Operational Safety in Nuclear Power Plants*. INSAG Series 13. International Atomic Energy Agency, 1999.

[66] U.S. Nuclear Regulatory Commission. *Reactor Oversight Process*. NUREG-1649. USNRC, Rockville, MD, 2006.

and workers' ability to raise safety issues, and finding and fixing problems. The Nuclear Regulatory Commission recently published a more definitive safety culture policy statement in the *Federal Register*. This statement includes a definition of safety culture and enumerates nine traits of a positive safety culture. Nuclear safety culture was defined as "the core values and behaviors resulting from a collective commitment of leaders and individuals to emphasize safety over competing goals to ensure protection of people and the environment."[67] The nine traits were (1) leadership safety values and actions, (2) problem identification and resolution, (3) personal accountability, (4) work processes, (5) continuous learning, (6) environment for raising concerns, (7) effective safety communication, (8) respectful work environment, and (9) questioning attitude. Some have criticized the early discussions of safety culture in the nuclear industry for being too narrowly focused on administrative procedures and individual attitudes at the expense of broader organizational considerations.[68] This most recent statement seems generally consistent with current thinking on safety culture.

HOW DO INSTITUTIONS CHANGE?

Most organizational change efforts occur in response to some type of failure or poor performance.[69,70] It follows that organizations seeking to change their safety culture are often doing so because of some significant safety-related problem or perceived vulnerability. In some instances, the actual problem may have occurred elsewhere, but the visibility and notoriety were such that other organizations were prompted or called upon to examine and reassess their own vulnerabilities.

Schein[71] presents a general culture change model that builds on the three basic steps or phases of Lewin's 1951 classic change model.[72] Schein describes the three stages as follows: (1) unfreezing and creating the motivation for change; (2) learning new concepts and new meanings for old concepts; and (3) refreezing or internalizing new concepts, meanings,

---

[67] U.S. Nuclear Regulatory Commission. Final safety culture policy statement. *Federal Register* June 14, 2011; 76(114): 34773-34778.

[68] Pidgeon, N., and M. O'Leary. Man-made disasters: Why technology and organizations (sometimes) fail. *Safety Science* 2000; 34(1): 15-30.

[69] Dunphy, D. Organizational change in corporate settings. *Human Relations* 1996; 49(5): 541-552.

[70] Weick, K. E., and R. E. Quinn. Organizational change and development. *Annual Review of Psychology* 1999; 50(1): 361-386.

[71] Schein, E. H. *Organizational Culture and Leadership*. John Wiley & Sons, Hoboken, NJ, 2010.

[72] Lewin, K. *Field Theory in Social Science*. Harper & Row, New York, 1951.

and standards. These stages reflect the fact that change involves unlearning as well as relearning. In essence, planned organizational change is a conscious learning process.

In the first phase, Schein emphasizes the importance of presenting enough disconfirming data to cause people to be uncomfortable with the current state. Moreover, these data should be linked to important organizational goals and ideals. The free and open exchange of information is a key attribute of a positive safety culture; it is also an important aspect of successful culture change. However, these disconfirming data, although valuable and useful, are really more symptomatic than diagnostic. At this point, further work is needed to take a detailed look at current safety systems, practices, and accountabilities to identify needs and set priorities. A multilevel systems perspective or HSI perspective can be useful to capture both the human and technical aspects of the work situation. Schein argues that change goals should be defined in concrete terms about specific problems that need to be solved and not as "culture change" per se.

Much of the success of the change process involves the creation of psychological safety. People need to feel secure and supported as the change and learning process proceeds. Unfortunately, too often, employees are viewed simply as passive recipients of change activities and other new initiatives.[73] Employee involvement can improve the fit and acceptance of new policies, practices, and routines by creating a sense of ownership and procedural fairness. From a culture-change perspective, involvement practices can help produce a push-pull situation where support for change is generated from both the top and the bottom of the organization. However, some organizational research has shown that employees are not always automatically ready to participate at the levels required, and efforts may be needed to build capacity in order to achieve the level of participation desired.[74,75,76] Indeed, the perceived lack of psychological safety can easily create anxiety and resistance among employees concerning anticipated changes, discourage them from participating, and ultimately defeat the entire change process.

The second stage of the change process focuses on learning and behavior change. Desired new behaviors can be coerced temporarily through

---

[73] Nielsen, K., T. W. Taris, and T. Cox. The future of organizational interventions: Addressing the challenges of today's organizations. *Work & Stress* 2010; 24(3): 219-233.

[74] Id.

[75] DeJoy, D. M., M. G. Wilson, R. J. Vandenberg, A. L. McGrath-Higgins, and C. S. Griffin-Blake. Assessing the impact of healthy work organization intervention. *Journal of Occupational and Organizational Psychology* 2010; 83(1): 139-165.

[76] LaMontagne, A. D., T. Keegel, A. M. Louie, A. Ostry, and P. A. Landsbergis. A systematic review of the job-stress intervention evaluation literature, 1990–2005. *International Journal of Occupational and Environmental Health* 2007; 13(3): 268-280.

the use of various enforcement protocols, but these behaviors are not likely to last if they are not accompanied by cognitive restructuring. The goal here is to change how people think about safety in their workplace, to change group norms, and reshape employee behavior-outcome expectations. Changing values, norms, and expectations is the essence of culture change. This is not easily accomplished through any single strategy or action. Consistent top management expectations and support are very important, but this change process almost always requires a well-executed, multicomponent plan that involves consistent messages through multiple channels, well-designed training activities, employee involvement, new methods and standards of evaluation, investment in new equipment and systems, and the use of role models or program champions. Changing safety culture involves altering the process of social exchange between employees and the organization. Social exchange theory[77] basically argues that employees evaluate their treatment by the organization and respond proportionally; this notion of reciprocity has been applied to workplace safety.[78,79,80] When managers and supervisors demonstrate their commitment and support for safety, employees reciprocate by expending greater effort to follow safe work practices and other safety recommendations.

To a considerable extent, the refreezing or internalization stage needs to show members that the new policies, programs, and behaviors are important and do produce the desired results.[81] Consistent with the learning perspective, this is a process of reinforcement and strengthening. The sharing of relevant information about safety performance is important, but even more important is showing that safety goals can be achieved without compromising other important outputs. Of course, the best situation is being able to show that improving safety actually improves other valued outputs. At this point, safety culture surveys, success stories, and employee interviews can be used to help sustain and reinforce the change process and provide additional evaluative data. Andrew Hopkins argues that where safety is a top priority, "the organization will aim to assemble

---

[77] Blau, P. M. *Exchange and Power in Social Life*. Transaction, Piscataway, NJ, 1964.

[78] DeJoy, D. M., L. J. Della, R. J. Vandenberg, and M. G. Wilson. Making work safer: Testing a model of social exchange and safety management. *Journal of Safety Research* 2010; 41(2): 163-171.

[79] Mearns, K. J., and T. Reader. Organizational support and safety outcomes: An uninvestigated relationship? *Safety Science* 2008; 46(3): 388-397.

[80] Neal, A., and M. A. Griffin. Safety climate and safety at work. *The Psychology of Workplace Safety*, J. Barling and M. R. Frone, eds. American Psychological Association, Washington, DC, 2004.

[81] Schein, E. H. *Organizational Culture and Leadership*. John Wiley & Sons, Hoboken, NJ, 2010.

as much relevant information as possible, circulate it, analyze it, and apply it."[82]

SAFETY SYSTEMS AND CULTURES

Much of the knowledge and experiences in the development of strong safety cultures in other areas can be transferred to academic chemistry research labs. Industrial research facilities, aviation, health care, nuclear power generation, and process safety all provide important examples of best practices that can be applied to all high-risk activities. The development of strong safety cultures in these fields demonstrates that training and reporting, peer communication, and hazard assessment are all key elements of a strong safety culture in any environment. For any of these practices to be adopted, however, organizational change must take place. To do this, one must understand the details and dynamics of the institution, the subject of the next chapter.

[82] Hopkins, A. Studying organisational cultures and their effects on safety. *Safety Science* 2006; 44(10): 875-889.

3

Laboratory Safety in Chemical Research in Academic Settings

INTRODUCTION

Chemistry and research with chemicals in university laboratories have been going on for centuries. Discoveries from chemical research carried out in university laboratories have led to revolutionary developments and advances in all aspects of the human condition. However, the key characteristics of colleges and universities, such as their diversity, "horizontal" decision structures, and tradition of faculty autonomy, present unique challenges for attempts to develop an institutional safety culture. This chapter focuses on the identification and explanation of the current status of issues and conditions associated with chemical safety and chemical safety management in today's academic research laboratories.

The organizational hierarchy and the responsibility for oversight of safety in university research are crucial elements in the development of a robust safety culture. However, determining who holds responsibility, authority, or accountability for the conduct of safe science in academic research institutions is often much more difficult than in non-academic or industrial settings. To ensure consistent, institutional involvement in establishing and maintaining a strong, positive, laboratory safety culture, participation in promoting safety must be encouraged at all levels, including members of senior university administration, provost and college and school deans, research administrators, environmental health and safety (EHS), department chairs, faculty and principal investigators, and lab researchers. Eliminating this current lack of clarity and consistency about safety roles and responsibilities across the university, particularly among faculty, researchers, and EHS personnel, is critical.

Variability in the regulatory oversight provided by federal agencies or state agencies, including state public universities, can also be a problem. Students and faculty from schools with little oversight are often caught off guard when moving to another institution where significant controls are in place. This issue is often compounded by a lack of standardized training of new faculty and students arriving at new institutions with varied external and internal oversight of safety.

Other challenges contributing to the existing academic laboratory research safety culture are numerous and include not only issues within the organizational hierarchy, but also physical limitations, such as problems with existing laboratory space and constraints on the design and construction of new research facilities. The increasing emphasis on multidisciplinary and interdepartmental research is a factor that needs to be carefully considered. Differing safety expectations in diverse areas of chemical research can be problematic.

A closer examination of the interface between the research laboratory and its direct leadership and support is a necessary step in promoting cultural change within the academic community. This core element of a strong, positive safety culture has not been developed in depth in other reviews; however, an understanding of the specific interactions, needs, and attributes of entities that are in direct contact with the research bench itself—the faculty/principal investigator, lab researchers, and EHS—is critical to development of sustainable change in academic research safety culture.

LABORATORY RESEARCH SAFETY

What Is Laboratory Safety?

An optimal laboratory safety environment would ensure that researchers setting foot in an academic laboratory, from inexperienced students to senior principal investigators, understand that they are entering a research environment that requires special precautions. It requires that researchers are aware of the hazards of the materials and processes that they and others in the lab are working with and are prepared to take rapid and appropriate measures to protect themselves and their co-workers, especially in the case of unexpected events. At a minimum, laboratory safety includes (1) awareness of the physical and chemical properties of laboratory reagents being used and of the safety and health hazards they pose; (2) availability and use of the proper apparatus and control infrastructure to carry out procedures safely; (3) knowledge and application of any additional special practices necessary to reduce risks; (4) familiarity and skill with emergency procedures including the use of safety showers, fire extinguishers, and eye stations; (5) a well-designed and organized workspace that facilitates safe operation, protects workers from hazardous environments, allows unrestricted movement about the laboratory, and allows for the segregation of hazards; and (6) use of proper personal protective equipment. In an ideal safety culture, all laboratory workers, including their leaders up to the highest levels of the organization, will naturally place highest priority on these practices.

The recent incidents have prompted academic faculty, staff, and administrators to ask two critical questions: *What will it take for us to educate ourselves and our students about the risks of our work and about the safety practices that allow each individual to make informed and aware decisions when carrying out research?* And, *if we are unable or unwilling to commit resources and personnel to provide students and researchers with competencies to handle the risks that accompany their work, should we continue laboratory work that involves the use of potentially hazardous chemicals?*

Most academic institutions strive to provide researchers with basic safety training and information, through interactions with the laboratory principal investigator, departmental safety coordinator, and/or university EHS staff. However, existing safety training programs often consist of lists of generic rules and regulatory requirements. Such requirements certainly merit discussion, but training that focuses on rules and regulations may promote a *culture of compliance* in academia, rather than a more desirable *culture of safety*. Evidence from other domains reviewed in Chapter 2 suggests that an effective way to promote a culture of safety in academic laboratories is to change the current training paradigm to incorporate not only regulatory awareness, but also in-depth work with safety concepts and practices that are central to research in the individual laboratory. Research practices that incorporate explicit analysis of the hazards and risks of planned work into research proposals and publications may promote better laboratory safety by preparing researchers to plan experiments with a critical assessment of and preparation for unexpected and potentially dangerous situations.

Faculty may not realize how little their students may actually know about the risks of a research laboratory and may simply assume adequate prior training. Both entering and experienced students may not know how to appropriately assess the risks of what they are doing, how to appropriately assess changes in risks if a key experimental parameter is changed, or how to keep a small error from getting out of control. Moreover, they may not realize that a process they used in the past without apparent incident was out of the ordinary, unsafe, or dangerous. Students, postdoctoral researchers, and their principal investigators also may not appreciate how rivalries, time pressures, and the emphasis on productivity can influence judgment and behavior.

Most, if not all, academic institutions that conduct chemical experiments have resources in place that can improve safety awareness and practices, but presentations to the committee suggested that many do not appear to combine them in ways that teach students core practices of chemical safety or that encourage self-aware behaviors in research laboratories. Some current practices may encourage faculty and students to view safety practices as prescriptive, bureaucratic annoyances that comply with requirements imposed by an external authority, rather than as practices that enhance safety and help ensure the progress of research.

There is wide agreement that protecting students and principal investigators is of primary importance and that, at present the academic community lacks a clear, unified vision about what a culture of safety entails. This stands in contrast to the apparent safety cultures that have developed in industrial research, in which everyone, from the CEO to hourly workers, understands and appreciates the relevance of safety to the mission of the company.

There are many different perceptions of the roles and responsibilities of those in the academic community, depending on where a particular person resides in the hierarchy of the institution. Various parties have often reported confusion or lack of information about the specific roles of other "players" and how these roles are interconnected.

CHARACTERISTICS OF UNIVERSITY-BASED RESEARCH ORGANIZATIONS

College and university organizations vary in many aspects, but most share some common characteristics that affect the focus, attention, and oversight provided for laboratory safety and the factors that contribute to their safety cultures.

Three key characteristics of colleges and universities are their diversity, horizontal decision structures, and tradition of faculty autonomy. Unlike business, medical, government, or military organizations with defined vertical structures, academic institutions are relatively flat organizations. The leader of an academic institution (often called the president or chancellor), the leader of the academic side of the institution (often called the provost), the deans of the colleges, and the chairs of academic departments or divisions may share more job characteristics with mayors or city managers than with business CEOs or chiefs of hospitals. From this perspective, one useful business analogy for the faculty or principal investigator may be the small business owner. Both are responsible for every function of their business and neither answers directly to their boss about safety. Just as the small business owner cannot leave hiring to a (nonexistent) human resources office or sweeping to the (also nonexistent) after-hours custodial service, the principal investigator cannot leave lab organization and cleanliness to the campus janitorial service or safety to the EHS staff. Just as the mayor or city manager does not order business owners to adopt fixed safety practices, but rather relies on inspections, fines, or (rarely) closures to provide business owners with incentives to maintain safe workplaces, so too do academic institutions rely on EHS surveys to provide the faculty with information, tools, and facilities to guide their safety practices. These academic incentives may need attention and incorporation of better practices to be more effective and to help promote and advance safety culture in laboratory research.

The tradition of faculty autonomy requires special mention. In U.S. academic institutions, individual colleges within a university, departments within a college, and faculty within a department have substantial autonomy over their research directions and practices. Faculty, working as individuals or groups, must seek and obtain a substantial part of the financial resources necessary to conduct research from sponsors outside the institution. A strong, positive safety culture must become an integral feature of this autonomy in academic chemistry laboratories.

Facility Characteristics

As noted above, a college or university site is more like a small city than a business or governmental operation. Most have large, dependent residential young adult populations living on site. Larger university entities sometimes operate their own power, water, and other utility systems. Some run public transportation systems for the campus and surrounding areas, operate their own police and fire response programs, manage large residential and dining complexes, and host and manage many large fine arts and athletic events on site, some attracting over 100,000 people to such events on the campus. In addition, colleges and universities are often visible political targets for local, regional, or even national issues.

Research colleges and universities often have several diverse laboratory teaching and research facilities. Although there has been a recent increase in the construction of newer research buildings throughout the sector, academic research facilities vary significantly in age and design. Older lab research facilities may lack modern engineering controls appropriate for the advanced research taking place in those facilities. The need for renovation and update of facilities and hazard control equipment to current requirements may be overlooked or considered lower priority by institutions and boards focused on new construction. Moreover, the costs for needed renovation and updating may be underestimated. In the current funding climate, principal investigators are unlikely to be able to fund the necessary safety-required facility upgrades. In some cases, funding agencies do not typically provide funding for safety upgrades to older facilities and do not allow direct grant funding for such expenses.

Newer research facilities may be designed with better engineering controls, but current designs that focus on efficient and flexible use of research spaces may contribute to overall higher risk to laboratory research occupants. For example, modern open-space laboratories that place the researcher desks and computer workspaces in close proximity to the research activity can be problematic because this approach places individual lab members who might be writing immediately adjacent to areas of chemicals use and storage. These unintended consequences of a well-intentioned design may increase risk to individuals within the laboratory. A safer arrangement provides for an office location outside the research activity environment for non-laboratory-based work. A particularly good arrangement separates desk areas from lab benches by impact- and fire-resistant glass, which protects researchers, but lets them monitor ongoing processes.

Organizational and Operational Structure

Within an academic institution, the research programs themselves are equally diverse. Modern chemical-use research ranges from basic science research in chemistry, physics, and biology, to applied research that crosses disciplines of engineering and medical sciences, to emerging sciences that span energy, nanomaterial, synthetic biology, and advanced materials. The diversity and scope of research conducted at academic institutions require a portfolio of approaches to establish and sustain strong safety cultures.

For example, researchers in engineering likely use different materials and processes than those working in medicine, synthetic organic chemistry, materials sciences, or a broad range of other areas. These differences in materials and processes can be accompanied by differences in hazards and risks, in safety training, and in safety culture. On occasion, these differences may hinder safety practices in collaborations. Indeed, different expectations about safety practices may create challenges for interdisciplinary collaborations not unlike those faced in corporate mergers between companies with distinct business cultures.[1,2]

---

[1] Weber, R. A., and C. F. Camerer. Cultural conflict and merger failure: An experimental approach. *Management Science* 2003; 49(4): 400-415.

[2] Bouwman, C. H. S. The role of corporate culture in mergers & acquisitions. *Mergers and Acquisitions: Practices, Performance and Perspectives*, E. Perrault, ed. NOVA Science, Hauppauge, NY, 2013.

Differences in research focus, tools, and chemical use are accompanied by a variety of management structures. Individual schools and departments or research centers may vary in organizational structure, based upon and reflective of the types of research conducted. Higher education organizations are often characterized by a flat structure with local authority and accountability, as opposed to the strong vertical hierarchy with strong authority and accountability within the management ladder, which is prevalent in industry and governmental laboratories where research is centrally funded and managed.

Populations

Another key characteristic of colleges and universities is the population served by and involved in academic research. Faculty members or principal investigators play a key role in fostering the safety culture and attitudes in laboratories. However, this role is not always emphasized or rewarded within the academic system and is often not modeled during graduate or postgraduate training. Even if such training is available, it is generally not standardized within an individual institution, much less across the research enterprise.

As the leader of the research laboratory, faculty members need to generate the research funding through increasingly competitive grant applications and awards. The faculty member also has to ensure and certify that the grant funding is managed and used properly in the conduct of the research activity and also comply with all the administrative work requirements of the grant agencies and host institution. A 2007 survey completed as part of the Federal Demonstration Project (FDP) Faculty Burden Survey concluded that

> [t]he data clearly show that the level of administrative burden is high enough to routinely take our nation's most qualified scientists away from their research. On average, faculty spent 42 percent of their time ensuring compliance with federal or institutional administrative requirements. Many of the associated processes do not fall within the faculty members' main areas of expertise, yet they are expected to be experts at managing issues related to affirmative action, accounting, and myriad other tasks. Meanwhile, given that multiple administrative tasks are spread out over each day, faculty members find it increasingly difficult to carve out the blocks of time needed to perform research and write about their results, or collaborate and adequately mentor their research trainees. Each year this problem becomes even more severe. In the FDP report, faculty members observed that the administrative burden has increased in recent years, which is not surprising, given the new regulations related

to homeland security as well as new attention to and requirements for financial accountability.[3]

The FDP repeated the Faculty Burden Survey in 2012 and found a similar outcome. Funded "researchers still report spending less than 60% of their research time actually engaged in research." The very nature of academic research—the pursuit of new knowledge—also engenders an entrepreneurial spirit, a part of which can resist central dictates or "one-size-fits-all" mandates.

Research populations in academic research labs involve relatively young individuals with limited experience, which is why such individuals are involved in academic research—to gain research experience. These young learners encompass a wide variety of research positions including research associates, technicians, postdoctoral fellows, graduate and undergraduate students, rotation students, visiting scientists, etc. For many of these individuals, the academic research environment is often their first research "job" in the laboratory, one they enter with little or no independent research experience but with a youthful exuberance. They are concerned about their future and about the impact of their attitudes on their adviser's opinion of them. With this concern, group members may avoid asking questions or engaging others in discussions about laboratory safety. Because of the nature of academic research laboratories as a training ground for new researchers in academic programs, there is a significant turnover of the laboratory research population. Such a high turnover rate in the core research population can make attempts to sustain a higher-level safety culture especially challenging and difficult.

Graduate students conducting research in U.S. academic research laboratories also increasingly come from diverse cultural backgrounds. In chemistry and engineering disciplines, international graduate and postdoctoral students may comprise 40 to 70 percent of the graduate researcher populations.[4] Some international students arrive with limited English skills and safety compliance knowledge, often with attitudes, practices, and values different from those in U.S. laboratories. Visiting scientists from all parts of the world also often carry out research in academic partnerships with U.S. researchers. In addition to different cultural backgrounds, visiting professors also bring their own safety culture and

---

[3] Decker, R. S., L. Wimsatt, A. G. Trice, and J. A. Konstan. A *Profile of Federal-Grant Administrative Burden Among Federal Demonstration Partnership Faculty: A Report of the Faculty Standing Committee of the Federal Demonstration Partnership*. 2007. Available at http://www.iscintelligence.com/archivos_subidos/usfacultyburden_5.pdf.

[4] Faculty Workload Survey (FWS). Preliminary Result Slides. Available at http://sites.nationalacademies.org/PGA/fdp/PGA_055749. Accessed March 12, 2014.

BOX 3-1
Student Rotations in Academia

In past years, graduate students entering most chemistry departments would participate in office interviews with faculty members during the process of choosing a research laboratory and an initial thesis project. In some departments, these face-to-face meetings were preceded by overview talks given by faculty members to the first-year students as a group. In recent years, this process has changed in a growing number of departments to a "rotation" system whereby each first-year student selects the (usually three) research mentors he or she is interested in working with, and then spends time (anywhere from 3 weeks to 3 months) actually working in each of those laboratories prior to joining one of them as a permanent member. This method of laboratory selection has been driven to a large extent by the fact that NIH training grants now require rotations for the first-year students supported by each grant. The rationale for this requirement may be that a longer-term exposure gives the students, principal investigator, and group members a chance to get to know each other better and thus make a more well-considered decision about which group to join on a permanent basis.

In other parts of this report, we have stated our agreement with departments that require general safety training for all incoming research workers, including first-year students. However, if a department has a rotation requirement for entering students, and these students are expected to carry out experiments during each rotation period, this raises additional safety questions. While there is likely to be overlap between the lab-centric training required to work safely in a particular group, there are also likely to be differences between the labs a particular student rotates through, as well as issues not covered in any general safety training for the entire first-year cohort. Since each student rotates through a different series of laboratories, and each laboratory is likely to encounter several new students doing experiments in their lab space, substantial individualized training is required to operate such a system safely. It is important that departments recognize this challenge and find ways to address it.

practices, for better or for worse, to the group that they are visiting (Box 3-1).

Academic research populations are also characterized by high levels of external and internal stressors. As mentioned above, the degree of intensity and competitiveness of chemistry departments can have a strong effect on a unit's willingness to embrace a strong positive safety culture. The impact of competition can be amplified by the additional stress created by competing deadlines, funding and publication demands, degree milestones, and personal circumstances. The level of stress faced by principal investigators and researchers can be a serious impediment to the practice of safe discipline in carrying out scientific research and, in some cases, may overwhelm an individual's capacity to function safely in the laboratory. In such cases, it may be important for faculty and researchers to make use of campus personnel or mental health resources.

ROLES, RESPONSIBILITIES, AUTHORITIES, AND ACCOUNTABILITY FOR THE CONDUCT OF SAFE SCIENCE IN ACADEMIC RESEARCH INSTITUTIONS

The organizational hierarchy and responsibility for oversight of safety in university research has been identified by other reviews of academic research safety.[5,6,7] In general, the descriptions below reflect the organizational structure for management and oversight of safety in academic research.

Senior University Administration

Responsibility for safety rests with the leadership of the organization. In academia, this leadership is the president or chancellor of the institution,[8] with varying input and oversight from a board of regents or board of trustees. Institutional leaders are responsible not only for creating a safe environment, but also for promoting a culture of safety. As noted in the NRC's *Prudent Practices* report, "leadership by those in charge ensures that an effective safety program is embraced by all. Even a well-conceived safety program will be treated casually by researchers and others if it is neglected by top management."[9]

Common academic administrative structures may dilute the commitment that senior academic leadership makes to laboratory safety. In a common structure, the president or chancellor assigns development and management of safety programs jointly to multiple units, such as schools or colleges, risk management units, and/or EHS units that may have no common reporting line other than the president or chancellor. This can create difficulties in identifying exactly who is vested with the day-to-day management of laboratory safety and hamper clarity about roles, respon-

---

[5] National Research Council. *Biosafety in the Laboratory: Prudent Practices for the Handling and Disposal of Infectious Materials*. National Academy Press, Washington, DC, 1989.

[6] American Chemical Society Committee on Chemical Safety. *Creating Safety Cultures in Academic Institutions*. American Chemical Society, Washington, DC, 2012.

[7] U.S. Chemical Safety and Hazard Investigation Board. *Texas Tech University Laboratory Explosion: Case Study*. Case No. 2010-05-I-TX. Washington, DC, October 19, 2011.

[8] The senior leader at a university can vary depending on the university or university system. For example, the University of California system and University of Texas system use opposite definitions of chancellor and president.

[9] National Research Council. *Prudent Practices in the Laboratory: Handling and Management of Chemical Hazards, Updated Version*. The National Academies Press, Washington, DC, 2011: 2.

sibilities, authorities, and accountability of individuals and organizational units for laboratory safety programs in the institution.

Additionally, the ability of senior university administrative officers to maintain a continued focus on promoting and sustaining a strong, positive safety culture competes with myriad other important issues that institutional leaders must contend with on a daily basis. Rapidly changing priorities and expectations, coupled with the increasing pace of turnover in senior leadership,[10] require that leaders in this environment build a management team that shares clear expectations and partnerships across academic and administrative units to foster laboratory safety and an institutional safety culture.

Provosts and College and School Deans

Horizontal academic organizational hierarchies often lead to centralized institutional programs, such as compliance programs and safety programs. This centralization can present a challenge to implementation and management, as these programs rely on flat organizational structures that are also responsible for overseeing other diverse programs and reducing budgets.

At many institutions, a provost (or titular counterpart) is the chief academic officer. This individual reports directly to the president or chancellor and oversees the colleges and schools. She is usually drawn from the academic ranks and increasingly may be hired from another academic institution. She may or may not have experience working in or running an academic research laboratory.

College and school deans are charged with the management of programs for their respective areas. In institutions in which the chemistry department is housed in a multidisciplinary college or school, the dean may be drawn from a discipline quite different from chemistry. As with all senior managers, deans must manage diverse priorities and most must manage with existing limited or diminishing funding. At the same time, deans are often charged with expanding academic programs, and, increasingly, fundraising to support existing and new programs. Reporting to the dean are department chairs, through whom the dean manages the academic programs and processes, including personnel processes such as promotion and tenure, curricular processes, budget and facilities, and any safety and compliance programs housed within the college.

At many institutions, deans and associate deans are unlikely to have detailed knowledge about the research programs in their units or about

[10] King, J. *American College President 2012*. American Council on Education, Washington, DC, 2012.

the facilities and practices needed to conduct specialized laboratory research. They may learn about facilities, infrastructure, and personnel needs primarily as budget requests. They may not have experience working in or running an academic laboratory. Such differences may create challenges as deans seek to identify expectations about laboratory safety in their units.

Deans, along with provosts and faculty governance, often oversee the personnel processes regarding tenure and promotion. College-wide guidelines for tenure and promotion typically describe processes for documentation and evaluation of three areas of faculty performance: teaching, research, and service. It is not clear to what extent, if any, these guidelines incorporate activities in support of laboratory safety, or to what extent such activities are included in materials given to faculty in college-wide "tenure academies" or in guidance for faculty seeking promotion.

Research Administration

Research administration and management in higher education (typically overseen by a vice provost or vice chancellor for research, or comparable title) plays a critical role in supporting and sustaining a safety culture in research. As with senior leadership, safety programs and a strong safety culture compete with many other mandates. Chief among these may be the attraction and maintenance of research funding and creation of facilities for new research opportunities. Research administration offices are often charged with many diverse responsibilities, such as establishing research compliance, various regulatory mandates and programs including conflict of interest, scientific misconduct, export controls, human participants and animal subjects in research, biosafety, responsible conduct of research, and intellectual property rights.[11]

The contribution of research administration to an institution's safety culture in academic laboratories is also influenced by a reporting structure that may dilute accountability for safety. In many academic research organizations, the research, development, and compliance programs report to the head of the research organization, while the institutional safety programs, including laboratory safety, report through a different branch of the organization, often through the facilities or financial administration structures. This can lead to a lack of accountability among the safety line management, the facilities management, the academic and research management, and the faculty-led research programs within the laboratories themselves. This bifurcation of organizational reporting can also

---

[11] National Research Council. *On Being a Scientist: A Guide to Responsible Conduct in Research*. The National Academies Press, Washington, DC, 2009.

affect the promotion and furtherance of a safety culture throughout the whole organization. Organizational structure and reporting of the safety support programs need to be in alignment with academic purposes and objectives to provide the most appropriate organizational alignment for a sustainable laboratory safety culture.

Environmental Health and Safety

EHS programs are an important component of the management of safety in academic research as well and integral to the promotion of the organizational safety culture. EHS programs in higher education also must manage and address a multitude of safety issues and programs that are endemic to higher education organizations and operations, as described previously in the characteristics of academic research institutions.

The organizational placement of EHS programs within the institution is variable.[12] As previously discussed, EHS programs in academia often report through the administrative support structures that include direct reporting to facilities, finance, risk management, or business administration lines. Historically, this reporting structure grew from the initial work by small safety programs to focus on those areas of the operations where injuries were, and remain, most prevalent; in facilities, dining hall, residential, and other manual materials handling operations. In others, safety programs were developed as special technical needs were identified. Programs specifically in support of research safety, such as radiation safety and, more recently, biosafety and biosecurity programs, have grown in response to new regulations and mandates. At times, these special laboratory support programs were initially started within the research administration line. Many institutions have coalesced their specialty technical programs into the existing EHS program structure. A small, but growing group of universities are requiring EHS to report through the senior research management programs, typically at the vice provost/vice president or higher level, which better aligns the EHS programs within the academic management system and may allow better access to overall research management. However, this trend is not widespread, and ensuring appropriate organizational reporting of EHS should be included as part of any review of an organization's overall safety culture to ensure optimal effectiveness and alignment.[13]

---

[12] Aon Global Risk Consulting. *Safety Management Function – Organization and Responsibilities—An Aon Survey*. September 2011. Available at http://www.aon.com/risk-services/thought-leadership/survey_safety-management-report.jsp.

[13] American Chemical Society Committee on Chemical Safety. *Creating Safety Cultures in Academic Institutions*. American Chemical Society, Washington, DC, 2012.

There is often confusion over the role of EHS with respect to academic research laboratories. Expectations of this role appear to vary depending upon the view of different responsible parties, especially among faculty and laboratory researchers. Institutional management expects EHS to provide safety, compliance, and risk management oversight of all campus operations, as well as provide assurance that institutional risk is being appropriately identified and managed. In contrast, some faculty members and EHS staff believe that EHS's role is primarily to serve as a regulatory entity, acting in place of external agency inspectors (e.g., Occupational Safety and Health Administration [OSHA], Environmental Protection Agency, and related state agencies). Others believe that the primary EHS role is to assist the research practitioners themselves in being compliant with external regulations. Indeed, academic administrators often task EHS with the responsibility of campus-wide compliance with all environmental and occupational health and safety regulations. When EHS personnel are not able to provide expert assistance to researchers regarding a safe procedure involving a specific or technical issue, a lack of respect ensues, and a confrontational relationship can develop.

Given this context, it is perhaps not surprising that faculty, postdocs, and graduate students are often confused as to the role of EHS relative to laboratory safety. For example, from the student perspective, EHS staff may be the ones who talk with students about how chemicals are stored or what types of shoes and goggles are needed. If EHS staff are the only, or the primary, people with direct laboratory contact with the students to talk about safety, a reasonable interpretation is that, "the people responsible for safety are the staff from EHS." Many EHS programs have professional staff able to consult on laboratory safety, but laboratory researchers should understand that EHS does not necessarily have, and in most cases cannot be expected to have, the same level and depth of focused technical skills needed to address the many diverse technical science research projects that take place concurrently in academic research on a campus. This lack of clarity and understanding of the role and authority of EHS can lead to negative attitudes on the part of faculty, graduate students and postdoctoral fellows, as well as cloud the roles, responsibilities, and accountabilities for safety within the academic research laboratory.

Department Chairs

Except for individual research faculty, department chairs in academia are closest in academic hierarchy to the actual conduct of research. It appears, however, that the assigned responsibilities of chairs rarely include an explicit mention of safety culture, and departmental processes and practices may not provide clear guidance about the chair's role and/or authority relative to laboratory safety. As with college and upper administration, competing priorities and lack of clarity over roles may reduce the likelihood that chairs assume or accept responsibility for safety.[14]

Departmental chairs in academia are typically interim appointments. They are generally senior faculty who rotate every 3 to 5 years through the administrative role of departmental chair while maintaining their own academic and research programs and interests or individuals who serve as chair and then move to other academic administrative roles. In some institutions, chairs are elected for specific terms (usually 3 to 5 years) by the faculty. As such, some, perhaps most, department chairs will circle back to being research faculty after their tenure as chair, and this places a high priority on maintaining a conflict-averse relationship with their peers. This can create climates in which chairs exert no clear authority to require, either on their own initiative or in accord with compliance or best-practice mandates from other institutional units, actions by other faculty or department members.

Chairs often oversee personnel processes regarding hiring, tenure, promotion, and faculty salary levels within the department. Position announcements for chemistry faculty describe requirements for evidence of promise in research and teaching, often with specific requirements for area of specialization and funding potential. Position postings may include quite specific expectations about publication and funding history but rarely, if ever, include expectations for safety experience or technical safety proficiency. It is not clear whether faculty interview procedures gather information about candidates' safety background, viewpoint, or abilities or the extent to which such information is considered in hiring decisions. When faculty members arrive in the department, practices associated with rearranging laboratory space and assigning space to the new faculty member vary widely. At some institutions, EHS and other personnel meet with faculty to ensure that the space and facilities are appropriate to the work planned and that the faculty member has the technical expertise to conduct the work and to train others to do so; at others, the new faculty member may only be issued a key and good wishes. A strong safety culture may make limited use of papers and grants as proxies for safety attitudes and actions, but rather be characterized by respectful inquisitiveness, by the chair or other senior faculty and by EHS, about a new faculty colleague's technical proficiencies and safety practices. Providing effective safety advice at this initial stage of a person's career has the strongest chance of inculcating strong safety culture in growing research groups.

---

[14] Id.

As is the case for college-wide guidelines discussed above, department-level guidelines for tenure and promotion usually describe processes for documentation and evaluation of three areas of faculty performance: teaching, research, and service. It is not clear to what extent, if any, these guidelines incorporate activities in support of laboratory safety, or to what extent such activities are included in materials given to faculty by departments or by their faculty mentors.

Principal Investigators

Principal investigators (academic research faculty members) play crucial and primary roles in laboratory safety and in development and maintenance of an effective safety culture within their research groups and within their departments. It is not clear, however, that the scope and importance of the faculty's roles are recognized and supported by all faculty members, or by their institutions. Academic research laboratories are operated quite independently from researcher to researcher. Principal investigators are expected to raise their own research funding through competitive grant processes, manage and oversee their awarded project grant portfolios, and perform other administrative duties that cannot be delegated to others. They are often not provided with management or mentorship training that is needed for the effective management of people.

Because of the need to regularly pursue grant funding and the administrative details related to managing funding, research faculty have less time to actually be present within their research laboratories on a regular basis. This may mean that they are unable to provide the necessary mentoring and direct management oversight, including safety oversight, to the research being conducted. This necessitates significant delegation from research faculty to postdocs and graduate students for the regular laboratory oversight and management responsibilities, including for safety within the laboratory, often without proper instruction, training, or authority. In some laboratories, the faculty principal investigator may no longer have the technical knowledge to set up or perform some newer procedures—especially if those procedures were developed after she or he completed training or, as often occurs in interdisciplinary research and emerging fields, has come from a different research discipline. In such cases, delegation is accompanied by the need to manage a process in which one is not always the expert.

Specialized safety training, specifically for faculty, is very limited and variable in content. Faculty are sometimes unclear about, or unaware of, the safety hierarchy and their individual responsibilities relative to laboratory safety, pointing to the graduate students, postdoctoral fellows, or the institutional EHS program as the personnel responsible for safety in their research programs. A university's expectation of the responsibility of faculty members for safety in their own laboratory research programs is often not made clear. The role that the faculty member has in providing leadership and setting the stage for promoting and advancing the laboratory research safety culture is often absent from many research groups.

Studies of safety cultures in other types of organizations suggest that perceptions about institutional commitment to safety play a significant role in faculty actions. For example, if the faculty perceive that colleagues do not discuss safety with their students or support consequences in the event of unsafe actions, they are less likely to engage in such discussions themselves. If they perceive that chairs, deans, or other university administrators will not cover the costs of mandated or recommended facility safety or environmental activities, requiring instead that funds come from direct grant dollars, which are not allowable under many awards, they may disregard the same administrators' safety exhortations. If faculty perceive that deans, chairs, and colleagues value grant income above all else when deciding raises, tenure and promotion, and award nominations, they will set their own priorities accordingly. Processes for annual faculty evaluations and tenure and promotion decisions provide perhaps the most visible criteria that faculty can use to judge their own efforts, and it appears that few faculty evaluation processes include opportunities and requirements for faculty to document their work to establish a robust safety culture in their laboratories.

Lab Researchers

Academic research program staff typically includes the following categories of personnel: research associates, postdoctoral fellows, doctoral students, master's students, undergraduate students, and from time to time, high school students and visiting scientists working on collaborative research. The majority of researchers in academic research laboratories are graduate students working in their first full-time research laboratory (perhaps after a modest amount of undergraduate research), along with postdoctoral fellows conducting independent research under the general direction of the faculty member with whom they are associated.

Several characteristics of these researchers may be critical to identifying the current level of safety culture in academic laboratories and to designing strategies to strengthen safety culture. First, most academic researchers are trainees. They are not permanent, long-term members of the laboratory, and their numbers and experience vary from person to person and fluctuate over time within a lab. They are at different stages of their educational and research training and may have different forms of financial support, or may even be paying for their graduate training. In a single laboratory, different trainees may work on quite different projects. They may have research deadlines that conflict with deadlines for academic courses or exams. They are often young, may or may not have support systems outside the laboratory, and are often encountering the complexities and pressures of academic research for the first time.

Entering laboratory research trainees differ in their experiences and expectations about laboratory work and in their knowledge about what it takes to conduct such work safely. Their college, or even pre-college, experiences may affect their expectations. In some school districts, hands-on high school laboratories, particularly in non-AP courses, have been replaced by demonstrations or online activities. Whether the changes in educational technology result from financial or personnel challenges, the disappearance of hands-on laboratories in science, technology, engineering, and mathematics disciplines has an influence on student skills and expectations. Students may arrive at their undergraduate chemistry labs with no experience in the special requirements of laboratory work and they may arrive with little awareness of the integral and important position of safety in laboratory research. Lack of awareness about safety and risk may also arise as an unintended consequence of changes in undergraduate chemistry laboratories. At some institutions, the undergraduate laboratories have been revised to focus on simplified, minimal-risk microscale experiments, limiting the numbers, types, concentrations, and amounts of chemicals used, the complexity of apparatus, and the variety of reactions. Such changes, pursued with valuable goals such as decreasing chemical waste, minimizing environmental impact, and reducing danger in laboratories containing large groups of beginning students, may decrease trainees' awareness, experimental experience (especially with larger-scale reactions), and understanding of and attention to questions of risk assessment and safety—and challenge faculty as they work to establish their research lab's safety culture.

Studies of safety cultures in other types of organizations suggest that perceptions about laboratory life might play a large role in trainees' actions. Trainees' perceptions of reward structures and expectations may contribute to a view that "time spent on safety is time not spent on my dissertation research." The committee heard experienced postdoctoral researchers and graduate students indicate that they feel disconnected from safety in their own laboratory. Although they may take online safety training, complete safety training quizzes, and so forth, safety practices are not consistent within or across research laboratories in a division, department, or institution. Moreover, trainees indicate that they do not feel empowered to address their concerns with others within the lab or with the faculty adviser. They also do not believe that they can move forward to effect positive safety changes without negative or punitive consequences to themselves. Students can feel uncomfortable confronting labmates and can feel that they do not have the power or authority to effect any changes without adverse negative consequences. Students also reported that the attitudes of principal investigators vary substantially among laboratories, and that this can affect how students approach safety in their own research. In addition, some have encountered students who do not follow the rules no matter how good the leader, and who may do so without consequences from their adviser or other leaders in the laboratory. These experiences lead to diminished value toward safety by the trainee.

In many laboratories, it is not clear whether hazard analyses of experimental procedures are being undertaken in any standardized form. Principal investigators focus on the science and research to be conducted, but it appears that not all investigators model or put priority on the need for a formalized identification of hazards inherent in materials and processes, or emphasize the need for a systematic and recorded risk assessment and safety plan (Figure 3-1). There appears to be a need for a more formalized approach to inclusion of hazard analysis, risk assessment, and safety as an integral part of the academic research process.

Additionally, it appears that for many laboratory researchers, formal safety education begins and ends with generic, and often online, safety training. While online materials or face-to-face lectures, and their associated assessments, can be effective ways to impart basic information about regulatory requirements and safe practices for laboratory work, they cannot substitute for engaging in the actions themselves. It appears that many current assessments of what researchers learn in safety training consists of written or online tests, rather than actions in a scenario in which the EHS professional and principal investigator set up a mock situation and say, "put these chemicals in storage," "clean up the spill," "is this apparatus ready to go?"

As stated in *On Being a Scientist*, "all researchers have had advisors; many are fortunate to have acquired mentors as well. An advisor oversees the conduct of research, offering guidance and advice on matters connected to research. A mentor—who may also be an advisor—takes a personal as well as a professional interest in the development of a researcher."[15] Appropriate mentoring by faculty, including a focus on safety in the conduct of science research, is a critical and primary element of promoting a safety culture in academic research.

[15] National Research Council. *On Being a Scientist: A Guide to Responsible Conduct in Research*. The National Academies Press, Washington, DC, 2009.

FIGURE 3-1 Complexities of student perceptions of where lab safety ranks. http://www.phdcomics.com/comics/archive.php?comicid=1613. Accessed November 6, 2013. Used with permission from "Piled Higher and Deeper" by Jorge Cham www.phdcomics.com.

EXTERNAL AND INTERNAL OVERSIGHT

There are numerous units that regularly inspect, evaluate, and advise on academic and management programs. Externally, these include regulatory programs such as federal or state OSHAs, granting agencies such as the National Science Foundation and the National Institutes of Health, and accreditation programs such as Association for Assessment and Accreditation for Laboratory Animal Care, that certify programs for adherence to professional standard of practice norms. Other accreditation bodies, such as the Association for the Accreditation of Human Research Protection Programs (AAHRPP), may also be used as an example.

External

Regulatory agencies vary from state to state in terms of what is expected and enforced. However, there are differences in whether regulatory agencies are effective or even have jurisdiction over some academic centers, depending on the type and location of the university or college. In some instances, federal agencies do not provide regulatory oversight for state agencies, including state public universities, while in others, such oversight by state and/or federal agencies is common. Students and faculty from schools with little oversight can be caught unaware when moving to another institution where such external oversight and internal controls are in place.

Most granting agencies do require that institutions receiving funding provide evidence of an active safety program, but do not require detailed information about the potential risks to researchers or safety of the specific proposed research as part of the individual grant application process. These agencies also do not necessarily provide oversight of laboratory chemical safety for grantees.

Professional associations such as the American Chemical Society (ACS) have been and are continuing to develop programs directed toward laboratory chemical safety. Still, the challenge is how to get this information disseminated to the appropriate parties and how to get people to use this information more effectively. These same ACS safety programs and guidelines have yet to be included in academic accreditation programs and thus are often not included in the academic training and instruction programs of the accredited institutions.

There are many accreditation programs for teaching and for research management, but these programs do not typically touch on issues of overall safety culture development in laboratories. By emphasizing a robust laboratory safety culture as critical for accreditation, the programs could provide additional support and incentive for enhancing and advancing safety culture at academic institutions.

Role of Funding Agencies

To date, funding agencies have relied on the institutions receiving grants to provide oversight. Those reviewing the scientific value of the proposed grant might be in a position to evaluate whether significant safety risks exist in conducting such research. Such review, however, does not currently include an assessment of whether the proposed grantee has the requisite knowledge of or understands the risks inherent in the proposed research. Identification or acknowledgment of the risk of such research is not typically part of either the grant proposal or the fund source evaluation process.

Some funding agencies may limit the use of direct grant funds and do not allow use of such funds for management of the safety risks of the proposed research. This is different for each granting agency and a policy that grant agencies should review and perhaps adjust to ensure better management within the laboratory of major grantees.

Role of Professional Associations and Publications

Graduate students, postdocs, and faculty should be more involved in setting safety rules and guidelines. EHS also plays a role, but one that may be more appropriate as advisory, as the primary responsibility for safety is with the researcher and faculty member. Principal investigators and researchers get their technical information primarily from peer-reviewed journals and other scientific association interactions. Journals and associations currently do not necessarily integrate science with the safety practices involved in the conduct of science to any great extent. When research is reported, there is seldom any remark about the safety precautions involved in carrying out the research activity that led to the desired outcome of good scientific data. There is a need for better integration of laboratory safety in the conduct of science, and journals and associations can play more of a role with such linkage of higher-risk research.

Internal

There are also a variety of internal institutional groups that provide review and audit research-related programs. Most relevant to the advancement of safety culture are the programs and oversight provided by the institution's EHS program.

However, EHS is not the only institutional oversight and auditing program available to review laboratory research. An organization's internal audit program periodically conducts management audits of various academic programs, primarily focused on financial and management systems auditing for compliance with myriad external requirements in those arenas. Internal audits seldom review management systems for the school or college's laboratory safety programs. However, some institutions have begun to include school and department management systems for safety and safety culture within the internal audit purview and review process. Some institutions have incorporated safety culture language into job descriptions and performance evaluations for all employees. These internal organizational approaches help to promote safety culture as a priority and serve as additional means to identify and support an awareness of safety as a core value for the institution, a key element for a strong safety culture.

Most universities carry out periodic audits of their various units, both academic and non-academic. A review typically involves the appointment of an external review committee composed of well-established and active research and teaching practitioners in the department's discipline from other universities, and sometimes includes individuals from non-academic institutions such as corporations. Typically, the department will draft a self-study document with contributions from various individuals, such as faculty, students, and staff. This is followed by an onsite visit by the review committee, during which an overall evaluation of the department's teaching and research is carried out. It is our perception, based primarily on reviews in which committee members have participated, that these exercises seldom involve analysis of safety culture and practices within the department.

CHALLENGES FOR EXISTING ACADEMIC LABORATORY RESEARCH SAFETY

New Research Facility Design

Laboratory space is among the most expensive to construct on a university campus; thus, it is understandable that experimental chemistry units and researchers try to maximize the amount of available hood and bench space. Unfortunately, cost restrictions often result in poor lab design from a safety point of view. For example, issues with desk space located in close proximity to working lab space is of concern for several reasons: First, for researchers to reach their desks, they have to pass through laboratory space requiring personal protective equipment, which is inconvenient at best and hazardous at worst. Second, individuals are concerned about their personal safety while working at their desks, particularly in cases where those desks are located close to another researcher's experimental work area as incidents occurring within other students' work areas could affect them. Finally, the lack of designated areas for students to eat, clearly situated away from chemically contaminated areas, is of concern. While National Fire Protection Association standards discourage this practice and sometimes require segmentation between hazardous and non-hazardous activities, the decision to segment is based on relative risk and is often complex. It would be helpful, perhaps, to have a national resource available that could provide reliable assessments, at the design stage, of the safety of new laboratories whose construction universities are considering.

Part of the job of educators is to train their students to do science in the "real world." That job is made even more difficult if laboratory space is not properly designed to ensure attention to safety. It is not surprising that industry recruiters often express concern at the lack of safety consciousness on the part of many newly minted Ph.D. graduates.

Multidisciplinary and Interdepartmental Research

Multidisciplinary and interdepartmental research is a significant area of growth in academic research. While this has led to exceptional advances, there is a risk that the increasingly interdisciplinary nature of research may lead non-chemists to undertake experiments involving chemicals without proper understanding of and training in the hazards involved.

It is not unusual to find projects that involve chemists working together with biochemists, cell biologists, engineers, and materials scientists, among others. Safety practices can vary widely from discipline to discipline. For example, researchers in biologically oriented laboratories that utilize chemicals do not always fully recognize the hazards of the materials they are working with. Although much of the discussion thus far has been focused around chemists working in chemistry laboratories, there are chemical hazards found in many other places on university campuses. Universities need to be cognizant that researchers in non-chemistry departments typically have less experience in the use of chemicals than many people working directly in chemistry labs. The lack of cross talk between disciplines concerning safety practices can lead to students undertaking experiments with no conception or little awareness of the risks and hazards involved.

SAFETY CULTURE KNOWLEDGE GAPS WITHIN ACADEMIC RESEARCH LABORATORIES

Hierarchical System Within Academia

As identified previously, a number of other reviews have focused on the academic research hierarchical systems and provided detailed recommendations for responsible parties outside the lab where the research takes place. These recommendations include a strong commitment from university leadership, including assurance of appropriate support resources, to sustaining a safety culture.

These recommendations are also very much aligned with this report's identification of the need for strong institutional support throughout the organizational structure and are reinforced in this document. The following list addresses some of the items identified as necessary to ensure a viable research safety culture:

1. Demonstration of safety as a core institutional value for the entire institution. This requires more than statements from leadership. It requires concrete demonstrations of how this value is prioritized and implemented throughout the organization.
2. Articulation of clear roles, responsibilities, authorities, and accountabilities for those directly involved in research safety within the laboratory, namely the faculty/principal investigator, laboratory researchers, and EHS staff that support lab safety.
3. Support for a strong, competent EHS program that is able to provide the technical support expertise necessary to maintain strong safety programs in research.

The existing hierarchical structure creates power differentials, impacts communication, limits upward feedback, and inhibits creativity and change. The current focus is on punitive outcomes or admonitions for focus on areas other than active research. There is a need for more focus and understanding regarding the elements at the research laboratory interface.

Research Laboratory Interface

A core element of a successful safety culture rests with the basic working group affected; for academic lab safety culture, this is the bench research group and its direct leadership and support. Understanding the specific interactions, needs, and attributes within entities that are directly in contact with the research bench itself is important.

The Venn diagram in Figure 3-2 illustrates three critical components of safety and safety culture within the research laboratories and the interdependence of these components in developing and advancing safety culture in academic research. What is needed is a better understanding of how these three major players can most effectively work together to advance the safety culture. Identifying the key attributes of advanced safety cultures in academic research labs and how each of the major players supports such advanced cultures will allow individual programs to better assess their existing programs and assign the roles, responsibilities, authorities, and accountabilities for laboratory safety culture advancement in academic research. The bottom line is that good science integrates safety directly within the research process and is valued by all direct and indirect participants.

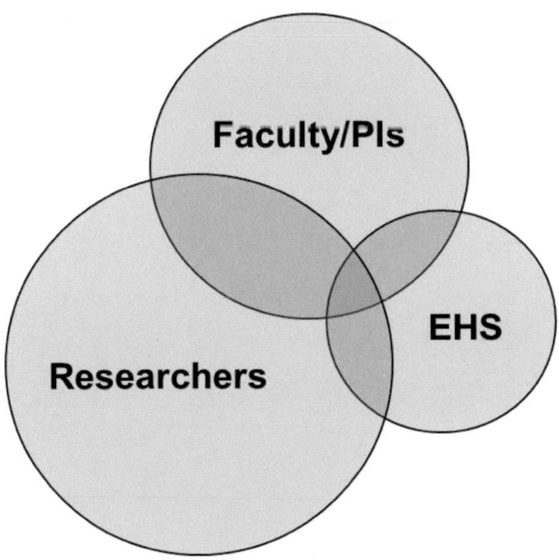

FIGURE 3-2 Lab safety culture at the bench top: Critical players and roles.

4

Laboratory Safety Dynamics to Improve Safety Culture

This chapter examines the interdependencies of the actors involved and the contextual features that make the academic research laboratories unique. Important among these features are the influence of personnel within the academic hierarchy, pressures for scientific productivity, feedback and communication channels, and the influences of external sources (e.g., funders, journals, and competitors). The chapter identifies well-recognized systems, lab processes, and practices that can improve safety performance in academic research labs. This coverage recognizes the complex and dynamic nature of the environment in which academic administrators, researchers, and students must work.

While large and small institutions have different resources to implement an effective culture of safety, it is also true that all institutions must meet certain safety requirements to operate and conduct scientific research. Positive safety performance is more difficult for some institutions to achieve given their resources, but none are absolved of the responsibility to provide a safe environment for their employees and students. Moreover, positive safety results can be an effective tool for recruiting and sustainability. Many of the same organizational processes, pressures, and practices apply to most academic organizations independent of their size.

Finally, while examples of practices from national laboratories are included in this discussion, we recognize that there are similarities and differences between these environments and the academic landscape. Nevertheless, organizations are encouraged to take advantage of lessons learned as good practices to be considered. Learning organizations take advantage of these successes and find ways to implement versions for their own purposes.

PRACTICES FROM NATIONAL LABORATORIES

Although there is still some debate on just how true it is that certain national laboratories have models for safety that go far beyond what is observed in a university setting, with some information in hand, it does seem true that academics can learn from these models and start to initiate their own. For example, a recent visit to a national lab (National Renewable Energy Laboratory) showed an impressively high level of safety precautions. From extremely safe and easy-to-use engineering controls in laser labs, to very high levels of documentation of chemicals and materials, this lab was a model for what many research labs should seek. Many of the procedures and precautions used in this lab can actually be found on the Web.[1] It is the desire of the Department of Energy to carry out precautions in a way that illuminates any possible weakness in their system regarding handing of chemicals or radiation exposure. The key safety personnel for each chemistry department can easily access this information and start to initiate their own departmental safety protocol.

INFLUENCES FROM THE TOP DOWN

A strong, positive safety culture instills thinking and behavior that assigns a high priority to safety. Such a culture encourages all concerned to have a questioning attitude about anything related to safety, to adopt a prudent approach to all aspects of their jobs, and to welcome open communications among different levels in the organization about safety issues. Chemistry laboratories are affected by hierarchies in the university, in the wider professional arena, in funding agencies, and in research organizational contexts. There are well-defined hierarchies within these entities that influence their ability to realize a vibrant safety culture. Several important factors influence this process on a variety of levels.

Academic Units

Chemistry departments house academic teaching and research functions and there are subunits within the department that can operate with a fair degree of autonomy. Department chairs, principal investigators, lab

---

[1] NREL: Environment, Health, Safety, and Quality. http://www.nrel.gov/ehsq/safety.html Accessed July 29, 2014.

managers, and graduate students head these units. Experience and anecdotal evidence support the description of the research units as academic "fiefdoms" where principal investigators have significant authority over their own research and operate autonomously as long as they do not intrude into other "fief" territories.[2] "Each fiefdom has an intellectual or administrative territory over which he or she reigns."[3]

There are hierarchies within these independent silos that can impede developing a culture of safety. First, the department head has administrative responsibility for safety in the department. The managerial responsibility of department chairs may conflict with their role as the principal investigator. Second, principal investigators may regard safety practices, such as inspections by outsiders, as a barrier to their research projects and violation of their academic freedom. Third, the individuals within the unit (lab managers, graduate students, and staff) are dependent, financially and educationally, upon a principal investigator's grant or research project. Taken together, these factors make it difficult to communicate safety concerns, raise awareness, or suggest changes.

Productivity as a Cultural Imperative

At the majority of U.S. institutions that conduct chemistry research, the faculty are expected to develop independent research programs and generate, from external sponsors, much, if not most, of the financial support necessary to support the equipment, supplies, and personnel, often including support for graduate students, required for research. As noted elsewhere,[4,5] these expectations and traditions of academic advancement create substantial pressure. Funding and publications are often given priority in decisions about advancement, salary, space, and other reputational issues. These pressures, combined with minimal if any training in personnel or laboratory management during the doctoral and postdoctoral periods or "on the job" in most universities, create challenges for the academic safety culture.

Within the hierarchy, graduate students' goals are aligned with these productivity goals because as one student succinctly captured it, "time is thesis." The more the researchers produce, the faster they can graduate.

---

[2] U.S. Chemical Safety and Hazard Investigation Board. *Texas Tech University Laboratory Explosion: Case Study*. Case No. 2010-05-I-TX. Washington, DC, October 19, 2011.

[3] Vangelisti, A. L., J. A. Daly, and G. W. Friedrich, eds. *Teaching Communication: Theory, Research, and Methods*. Routledge, New York, 2013.

[4] U.S. Chemical Safety and Hazard Investigation Board. *Texas Tech University Laboratory Explosion: Case Study*. Case No. 2010-05-I-TX. Washington, DC, October 19, 2011.

[5] American Chemical Society. *Advancing Graduate Education in the Chemical Sciences: Full Report of an ACS Presidential Commission*. American Chemical Society, Washington, DC, 2012.

There is a pressure to publish, but there is also the pressure to come up with results that the leader (professor, principal investigator) is seeking. This leads to quantitative workload stress, derived from the need to keep working to retain one's job and avoid getting "scooped" by a colleague or competitor. It also produces demands in terms of qualitative workload stress—that is, the need to keep working until you find the results you targeted in your research project. Finally, the power differences between the principal investigator and graduate students can inhibit the reporting of hazards, incidents, shortcuts, or near misses. This is relevant because of educational hurdles as well as keeping the funding for the research unit.

The pressures to produce results are further fueled by the fact that financial support for graduate students relies heavily on individual research grants. This reliance on grants to support students creates a potential conflict between a culture of safety and productive grant-supported research.[6] Decoupling graduate students' dependence on grants for financial support may provide a useful way to enhance the development of positive safety culture in research groups.

There is evidence that the social context that these productivity pressures create can cause injuries. External loads, organizational factors, and social contexts were hypothesized to have a relationship to repetitive strain injuries. Since then, there has been evidence that emotional and psychological demands can have effects on biomechanical functioning.[7] Injuries further erode the culture of safety within the unit.

YOUNGER PEOPLE AT WORK AND RISKY BEHAVIOR

Because of the composition of academic laboratories, it is important to make special mention of evidence that young people differ from more experienced researchers in their perceptions about risks that affect their behavior. A National Academies study examined how youth are different and are affected by the way that work is organized and managed, with possible generalization to postsecondary students.[8] While university students are not children or adolescents, there is certainly a range of maturity and development within the university community and some of these trends may be applicable.

---

[6] Id.

[7] Marras, W. S., K. G. Davis, C. A. Heaney, A. B. Maronitis, and W. G. Allread. The influence of psychosocial stress, gender, and personality on mechanical loading of the lumbar spine. *Spine* 2000; 25(23): 3045-3054.

[8] National Research Council. *Protecting Youth at Work: Health, Safety, and Development of Working Children and Adolescents in the United States.* National Academy Press, Washington, DC, 1998.

Type of Work

Young workers are often engaged in work with high turnover, little on-the-job training and limited discretion, uncertain hours, low pay, and few benefits. Jobs with these qualities tend to be more dangerous, and are often found in small businesses, much like a laboratory setting. Studies point to a negative relationship between an organization's size and risk of injury or death. Like small businesses, university labs may have high turnover, leaving more inexperienced workers in charge of potentially dangerous tasks. University labs are also more exposed to market pressures, which may lead them to ignore safety procedures by cutting corners. A National Institute of Occupational Safety and Health survey found that smaller organizations (fewer than 100 workers) provided less training, conducted fewer inspections, and used fewer professionals in their safety programs.

Risk Assessment

How young people recognize and assess risks and how they decide on which courses of action are important to all aspects of university life. As children develop into adults they begin to generate options, look at situations from different perspectives, anticipate consequences, and evaluate the credibility of sources. By mid-adolescence, young people can make decisions similar to those of adults.

There are data to indicate that injured teens may have taken on tasks to prove that they are responsible and independent. They performed these tasks despite knowing that they were dangerous or violated laws but acted in fear of losing their jobs. There may be analogies between behavior in these situations and in university laboratories.

Another report, *Improving the Health, Safety, and Well-Being of Young Adults: Workshop Summary*,[9] highlights the differences between younger workers and adults and the interventions that seem to be effective in improving health and safety:

- Young adults tend to have the lowest awareness of risk and the least access to health care and insurance.[10]
- Brief interventions, including skills-based interventions, motivational interviewing, and personalized normative feedback are

[9] National Research Council. *Improving the Health, Safety, and Well-Being of Young Adults: Workshop Summary*. The National Academies Press, Washington, DC, 2013.

[10] Id., p. 45.

effective methods for reducing risky behavior, such as drinking among college students.[11]

- Peer-to-peer interventions can achieve buy-in, trust, and rapport in creating effective change.[12]
- Rewarding those young people for good positive behavior, rather than punishing bad negative behavior, may achieve getting young people involved in reducing undesirable actions. For example, if someone is in trouble for drinking or drug use, the person offering the help should not get in trouble for reporting the problem. Young people can provide resources for their friends by becoming involved. They should not be punished for this reporting.[13]

COMMUNICATION ABOUT LAB SAFETY

Communication about lab safety is couched in the language of compliance. There is a stronger emphasis on compliance than on safety. Understandably, administrators are keenly aware of managing perceptions about organizational safety and its impact on the institution. This leads to the enactment of policies and procedures designed to mitigate these risks. This is often done as a top-down approach to creating change. At the same time, technical support staffs (including environmental health and safety [EHS] and chemical safety personnel) are familiar with mandated standards that must be met to comply with regulations. Professional staffs have a sense of urgency because they understand the technical aspects of the requirements and regulations and because of their genuine interest in mitigating risks to people. The actions they produce are often grounded in regulatory directives, or prohibitions to autonomously functioning individuals and research units.

Communication Content

Most of the measures reviewed from chemistry laboratories are *lagging indicators* of safety performance. That is, they record what has already occurred, tend to have a negative tone, and seem to be affixing blame. To change behavior and the culture, organizations should be monitoring *leading indicators*—measures that can prevent incidents and mitigate risks. Lagging indicators are more typical of a compliance-based, reactive approach. Typical lagging indicators would include parameters such as the number of accidents, incident rates, deaths, body part affected, time

[11] Id., p. 87.
[12] Id., p. 88.
[13] Id.

of injury, reasons why the injury occurred, profile of the injured worker, direct or worker compensation costs, and number of lost workdays.

Leading indicators could include, but are not limited to, near misses; lessons-learned databases; research group meetings focused on safety; job safety analyses completed and trends therein; surprise inspections and their results; case studies highlighting good practices; results of suggestion programs and changes made; training opportunities, requirements, and resources; awards for positive actions; behavioral observations completed; principal investigator coaching; intra-lab coaching and information sharing; and safety perceptions about how people throughout the organization view safety. These kinds of data highlight the importance of changing behavior and allow information to flow upward in a hierarchy. Moreover, if the leading indicators were to be tied to decisions such as promotion, salary increases, and resource allocation, they could influence peoples' behaviors in meaningful ways.

A systems approach is needed to manage any changes and to avoid serious injuries. A thorough analysis of risk in complex systems considers more than the technological and engineering solutions. It requires addressing the psychological, social organizational, and political processes that contribute to incidents.[14] One implication is to understand the leading indicators to change individual and organizational behavior. Human factors and ergonomics principles and systems safety have been used to change many complex systems using leading indicators.[15]

Although the context differs from industrial examples, the same principles can be applied to academic chemistry research laboratories. A proactive systems approach is needed to influence individual and organizational behavior. Forward-looking methodologies and metrics can avoid the unintentional blindness caused by a compliance-based approach.

Implementation

The top-down approach is often met with resistance, in part, because the policies and procedures may not seem to make sense, or have any real validity, or may be perceived as being at odds with research productivity. This is especially true when requirements are promulgated by those without any experience in a specific research area or when a policy or procedure is expected to cover a wide range of applications. Further, if the demand for action is perceived as a response to litigation or as a defensive

---

[14] Bea, R., I. Mitroff, D. Farber, Howard Foster, and K. H. Roberts. A new approach to risk: The implications of E3. *Risk Management* 2009; 11(1): 30-43.

[15] National Research Council. *Macondo Well Deepwater Horizon Blowout: Lessons for Improving Offshore Drilling Safety*. The National Academies Press, Washington, DC, 2011.

action, the approach may be seen as geared to match compliance demands rather than as an active attempt to improve safety culture. The negative reaction is both predictable and understandable.

A less than enthusiastic response can be expected when professional staff assert that the reason for doing something is because "it's the law." Moreover, when managers and responsible individuals are threatened by regulations, the modus operandi is to practice avoidance behavior rather than proactively seek positive outcomes. Finally, when policies and procedures establish minimum standards, these become the target ("satisficing").[16] Instead, a true culture of safety should involve optimizing conditions through desired behaviors.

A prime example of this was found with the University of California's response to its settlement with Cal/OSHA. It developed laboratory safety policies for Laboratory Safety Training, Personal Protective Equipment, and Minors in Laboratories and Shops. After its initial draft and much negative reaction from researchers, the policy had to be reworked to make reasonable accommodation for practical implementation by laboratories.

More than a set of standard operating procedures and policies, a culture of safety extends beyond departments to all members of the organizational community. This will require a campus-wide approach to changing the safety culture. Partnering with other labs, departments, and colleges can have a much higher synergistic effect than a single laboratory making changes in isolation.

The motivation for changing practices should be to improve the working conditions in laboratories to enhance the quality of research, protect its people, and create sustainable results. If done correctly, compliance will follow. Focusing on a compliance strategy alone has a less likely chance of developing a positive safety culture.

LEADERSHIP SHOULD INCLUDE SAFETY AS A VALUE AT ALL LEVELS

Leaders at all levels in the organization must demonstrate that safety is a value and must convey their expectations to their followers. Who are these leaders? Similar to other industries and organizations, "the ultimate responsibility for creating a safe environment and for encouraging a culture of safety rests with the leadership of the organization and its operating units."[17]

The investigation of the 2010 incident at Texas Tech revealed "safety

[16] March, J. G., and H. A. Simon. *Organizations.* Blackwell, Cambridge, MA, 1958.

[17] National Research Council. *Prudent Practices in the Laboratory: Handling and Management of Chemical Hazards, Updated Version.* The National Academies Press, Washington, DC, 2011.

policies either did not exist or were not enforced. No single person or entity within the university was accountable for ensuring that the CHP was up-to-date, enforced, and applicable to the laboratories it was meant to regulate."[18]

Often, researchers who manage projects are unaware that they are the persons responsible for safety in their organization. Clear lines of authority and responsibilities that come with positions should be articulated clearly to everyone.

Analysis of tragic events in complex systems[19] have shown that failures

> can be traced back to management processes that did not provide adequate controls over the uncertainty of human decision making, ... Management processes failed to adequately identify and mitigate risks created by operational decisions prior to the blow out, communicate critical information, train key engineering personnel, and ensure measures taken to save time and reduce costs did not adversely affect overall risk.[20]

The lesson learned here is that leadership needs to be exerted at all of these levels to create a culture of safety.

The U.S. Chemical Safety and Hazard Investigation Board (CSB) investigation points out that *there is no single point of failure in serious incidents. The event is the result of a complex interaction among diverse actors across levels of the organization.* While most accidents focus on the human error, or mistakes made by the person directly involved, deficiencies can be found throughout the organization that contributed through inaction, poorly defined roles or expectations, training, enforcement, and/or monitoring. Therefore, strong leadership should be taken at all levels of the academic institution. Moreover, leadership should address not only the technical and engineering aspects of safety, but also the psychological, social, organizational, and political processes involved in causing injury events.

---

[18] U.S. Chemical Safety and Hazard Investigation Board. *Texas Tech University Laboratory Explosion: Case Study.* Case No. 2010-05-I-TX. Washington, DC, October 19, 2011: 14.

[19] National Research Council. *Macondo Well Deepwater Horizon Blowout: Lessons for Improving Offshore Drilling Safety.* The National Academies Press, Washington, DC, 2011.

[20] Id., p. 76.

INFLUENCES FROM THE OUTSIDE IN

Incorporating Safety into Performance and Evaluation Measures for Faculty

The daily routine of most faculty members is filled with many responsibilities. These responsibilities range from educational activity and academic research to administering, planning, and executing new initiatives as well as departmental service (which includes teaching and committees). Generally, these responsibilities constitute the bulk of the evaluation of the annual success of each faculty. These are the core parameters for which promotion and tenure as well as merit-based salary standards are set. Thus, great attention is placed on these areas of activity in each department each year. This is already a substantial set of responsibilities, which indeed keeps faculty members who run research laboratories very busy. In a research leadership position, laboratory safety is also a major responsibility. However, the level of importance that is placed on laboratory safety in various chemical laboratories in reference to the overall evaluation of a faculty member's performance is not as certain. This leads to the question of how much a faculty member's safety practices should be weighed in considering advancement within a department.

The question of impact (reward) of a faculty member's safety practices is as much a matter of research and scientific discipline as it is a matter of culture. The first important issue to remember is that the need for laboratory safety is not only good for the health of the students and researchers involved but also in educating and providing a positive example to younger scientists that laboratory research can be done safely and, at the same time, efficiently. The practice of laboratory safety is ultimately left up to the individual, and in most cases the importance of doing research safely is learned from others in the same lab. A faculty member's leadership skills are truly tested in both illustrating the importance of lab safety and enforcing its practice at all times. There exists a temptation to sacrifice this responsibility, at times, out of a perceived need to conduct particular experiments when time and/or resources are limited. The faculty member's leadership and exemplary discipline in carrying out proper safety precautions is needed most in these situations. When safety precautions are neglected in the lab, it is the responsibility of the faculty member to use measures necessary to eliminate this behavior so as not to harm others in the lab. This is indeed a matter of scientific or research discipline. However, because the previous research experiences of each lab member may vary, so too will their level of discipline in safety behavior. Thus, there is a cultural aspect to the demonstration of a faculty member's attitude toward safety. While it is certain that each faculty member may have experienced varying cultural attitudes toward safety, it is now clear there is little room for this diversity in allowing bad safety habits to exist and ultimately to harm those who are present in the laboratory. Department practices that place real importance on safety during annual or advancement evaluations of each faculty member could have a large impact on changing the culture of safety in academic laboratories.

There are several reasons that would justify a department using laboratory safety as one measure of a faculty member's advancement. Perhaps the most important of these is that this could be a good preemptive strategy for preventing accidents or injuries. If this is a generally accepted practice and each faculty member is aware that his or her annual evaluation is partly dependent upon their safety practices and safety management, this may provide more uniform safety behavior (culture), which is safety culture in the department before an accident actually happens. Unannounced safety checks may also provide a good measure of each faculty member's performance in this regard. This would also be a good way for the department to evaluate the progress of each of its faculty members over time in providing a safe research environment. As mentioned above, while the discipline of performing research is dependent on the leadership of the faculty member and ultimately the individual doing the research, the possibility of changing the cultural attitude toward safety is also the responsibility of the department and research/university community. An additional reason is that including safety in annual and advancement evaluations allows faculty members to document and report the substantial work required to develop and sustain a strong, positive safety culture in their laboratories. It encourages faculty to measure and report leading indicators for their groups, as metrics of adaptation to rapidly changing research programs. Using more direct and formal methods of evaluating a faculty member's discipline, leadership, and, ultimately, cultural attitude toward doing laboratory research in a safe manner could make a difference in reducing the number of incidents each year.

Journals Should Include Safety and Health Information

Publication is a major component of academic life. As mentioned previously, it is also a major factor in promotion and tenure decisions. It is the driving force behind the hard work and effort of aspiring graduate students. Because this high level of ambition and enthusiasm may at times cause some scientists and engineers to make hasty decisions about safety, the publication process may also be used to define, describe, and defend the important safety precautions and practices necessary to carry out research. Some journals encourage the inclusion of safety information when particularly hazardous materials are used in experiments documented in an article. Since many experiments involve potentially hazardous procedures, making safety information a regular component of most or all experimental papers would provide a strong incentive to the development of more widespread safety culture.

The manner in which this could be enforced by particular journals (in chemistry) is relatively straightforward. In each publication of a full article (or even in communications), there is a section for experimental details. This section should be expanded to include strategies for hazard identification and risk mitigation. The purpose of this expansion is not only to inform future researchers about the hazards of carrying out a reported procedure, but also to allow the young scientists writing the papers to recognize that this is a professional requirement. Much of this can be formulated into procedures that many of the lab members can utilize in their own papers. In instances in which unanticipated hazards or risks are discovered during data acquisition or analysis, safety information must be included in results and discussion sections and in the abstract and any publicity about the work.

LABORATORY PROCESSES

Hazard Analysis

Hazard analysis involves the identification, assessment, and mitigation of hazards and their associated risks. It is a process to assess risks and ensure that those risks are mitigated or eliminated before initiating any laboratory work. These are critical skills for an individual to know and apply. One should assume that no activity is guaranteed to be absolutely risk-free, especially when some hazards may not have been identified, assessed, or properly mitigated. In addition, one cannot assume that hazards remain unchanged even on routine jobs or with any task requiring job hazard analysis.

For the hazard analysis to be successful, all individuals involved are required to participate and be able to recognize and identify hazards. Hazard recognition and identification can only be obtained through training and continuous feedback (e.g., during walk-throughs, observations, and peer-to-peer feedback). This learning process must be extended to all individuals involved in research: undergraduate and graduate students, postdocs, faculty/teachers, principal investigators, laboratory managers, coordinators, etc. To build a long-term, well-informed/educated culture of safety, this process should start at the undergraduate levels and be incorporated into academic research at all levels, including thesis and dissertation proposals, laboratory notebooks, presentations, and publications.

Laboratories

The designs associated with safe, efficient laboratories have evolved over time. In synthetic chemistry laboratories, two factors that have changed significantly are the ratio of hood to open bench space, and the relative locations of space in which active experimentation is going on and space in which writing, computations, and other desk work are being carried out. Academic laboratories built before 1950 had significant bench space but little associated fume hood space. It was common practice to carry out chemical reactions (even ones involving highly toxic chemicals) on laboratory benches where the researcher was not protected by a fume hood. Gradually, the ratio of hood to bench space increased as new buildings were constructed, but the common standard of one hood per researcher was not institutionalized in many laboratories until the past 10 or 20 years. There are still many laboratories in which the available hood space per researcher is limited, resulting in experimental procedures involving hazardous chemicals and gases are being carried out on benchtops or on vacuum lines situated outside of fume hoods.

Physical and biological laboratories raise potentially problematical issues. In years past, most physical chemistry groups performed relatively few syntheses. However, with increasing interest in novel and functional materials, such groups have been carrying out more synthetic work. The amount of fume hood space in typical physical chemistry laboratories, as well as the perception of the risks involved in carrying out synthetic procedures, is often too limited. Biological laboratories face a similar risk. Groups that work with highly toxic organisms or certain radioactive materials have special laboratories designed to protect workers from those hazards, and there appear to be good protocols and campus oversight for those activities.[21] However, it is clear that in many routine situations, many researchers in the biological sciences feel their experiments are free of chemical hazards, perhaps because they are performed mostly in aqueous media. This leads both principal investigators and researchers to believe that common chemistry laboratory safety practices, such as wearing safety glasses, lab coats, and protective footwear and gloves, are unnecessary—even in cases in which biological materials are being modified with potentially hazardous chemical reagents.

Many physical chemistry laboratories have an additional possible concern that may need to be addressed. This involves the use of lasers. This equipment raises the important issue of eye damage from accidental exposure of co-workers' eyes to laser irradiation. This can be prevented

[21] For more information on biosafety, including the biosafety level guidelines, see http://www.cdc.gov/biosafety. Accessed July 29, 2014.

by the rigorous installation of interlocks and, better yet, the installation of devices that allow the positioning of the elements used in laser experiments (e.g., mirrors, detectors, and spectrometers) by remote control, which minimizes the accidental exposure of the experimenter to laser beams (Box 4-1).

BOX 4-1
Laser Safety Anecdote

The use of Class 3 and 4 lasers in academic and other research institutions has become commonplace. There have been a number of serious accidents involving exposure to laser beams. These may result from a lack of training, experience, or safety culture for those involved, or possibly point to the need for critical, yet costly, engineering controls. These incidents have involved both new and experienced scientists and engineers. For example, one recent incident involved a graduate student and a visiting scientist with more than 15 years of laser experience. Both researchers were working with a Class 4 multiple laser system at full power when the scientist was struck in the right eye by specular reflection, resulting in a retinal burn and a loss of acuity in the eye. Neither researcher was wearing laser eye protection while repositioning a mirror element that investigators believe caused the beam to reflect off a stainless steel mounting post. Laser eyewear was not worn so that the researchers would see a small amount of visible light from the laser while aligning the mirror. This was a clear violation of standard operating procedures that specified the use of laser eye protection.[1]

A number of safety precautions were overlooked in this incident. For example, the potential for eye exposure while repositioning optical elements was not even considered during the work planning process. Even if it had been, would (or could) the incident have been prevented with the use of specific engineering controls? At a minimum, it is clear that lasers should be equipped with a protective housing, a clearly identified aperture, and a clearly marked switch to deactivate the laser or reduce its output to less than maximum permissible exposure. However, in this example, as is the case in a large number of the accidents involving laser exposure, the laser light came from a specular reflection, not directly from the laser beam. The use of engineering controls[2] is thus necessary to protect all individuals in a laser room, even those who are not actually performing the experiment on the laser table.

---

[1] United States Department of Energy. *Special Operations Report: Laser Safety.* U.S. Department of Energy, Washington, DC, February 2005. Available at http://jrm.phys.ksu.edu/Safety/DOE_Laser_Safety_Report-Mar-05.pdf.

[2] University of Waterloo Safety Office. *Engineering Controls.* Available at http://www.safetyoffice.uwaterloo.ca/hse/laser/documents/control__engineering.html. Accessed July 8. 2014.

Engineering Controls

Engineering controls, with complete elimination of a hazard, are at the top of the hierarchy for safe experimental design. A number of research institutions have used engineering controls to remove a hazard or place a barrier between the worker and the hazard. Well-designed engineering controls can be highly effective in protecting workers and will typically be independent of worker interactions to provide this high level of protection. The initial cost of engineering controls can be higher than the cost of administrative controls or personal protective equipment. This is especially true in dealing with engineering controls for electronic or laser equipment. However, over the longer term, operating costs are frequently lower, and in some instances can provide cost savings in other areas of the process.

Distribution of Costs

As noted earlier, many universities offer research workers the option of obtaining free prescription safety glasses. However, in some places the cost is charged to the principal investigator's research grants. Similarly, chemical waste disposal in many institutions is covered by university funds, but for others, the cost is also recharged to research grants, just as it is for safety glasses. Recharging safety glasses and hazardous waste disposal costs to grants incentivizes researchers to take shortcuts that could result in injury or damage to the environment.

Important Characteristics in the Laboratory

In a strong, positive safety culture, researchers are encouraged to care about working safely and are rewarded, rather than sanctioned, for this philosophy. One of the most recalcitrant problems in many chemistry laboratories is the attitude, unfortunately often reinforced by principal investigators, that safety practices are time-wasting inhibitions to research productivity. Efforts must be found to convince such people that working safely enhances, rather than inhibits, research productivity. Certainly, an accident is one of the most serious inhibitors of research productivity. Thus, one would think that principal investigators would have a strong incentive, for that reason as well as many others, to foster a positive safety culture in their laboratories.

Strong, positive safety cultures will develop when researchers care about and promote working safely, and when institutions have an obligation to monitor working conditions to ensure that they are safe and that the procedures being used are safe. The classical approach involved is enforcement, that is, strong sanctions for people who do not work safely.

Although this may be necessary in some cases, and is one of the factors that maintains safety in industrial research laboratories where people can be fired for safety violations, we believe that using only the "stick" rather than the "carrot" is not the most rational way to ensure a strong safety culture. We believe that encouragement and rewards for good safety practices are both more effective and result in a more collegial and safe university laboratory environment.

In this context, the following characteristics should be sought and encouraged in laboratory environments to ensure that laboratories have strong safety cultures:

1. Laboratory safety culture is strongly influenced by the extent to which research workers are consulted about safety rules and procedures. Rules handed down from the administration in the absence of such consultation tend to be designed in a one-size-fits-all manner, which may apply reasonably well to one type of research laboratory, but not very well to others. This not only creates inefficiencies, but also produces hard feelings on the part of research workers, which can erode any hope of developing a culture that encourages researchers to care about working safely.
2. There are facets of a rational award structure that can be improved in most universities. One important target should be the group meetings that almost all research groups hold on a weekly or other regular basis. If incentives could be found to devote some period of time every week to safety issues at these local meetings, it would go a long way toward the establishment of a positive safety culture in specific laboratories.
3. Funding agencies may choose to include these factors in grant evaluations, for example, as part of the "broader impacts" sections that are now being required in National Science Foundation grant proposals.
4. Even if a reward structure for working safely can be developed, administrators have an obligation to make sure that proper safety procedures are being followed in their institutions. Most chemistry departments (or at least universities) have one or more safety officers who are responsible for monitoring laboratory environments and working out ways to deal with problems that arise. The existence of such positions is important, but note that in many institutions the safety officer is overwhelmed by the large number of laboratories that he or she is responsible for, and especially by the diversity of activities (e.g., synthetic work, laser experiments, biological studies) subject to monitoring. To alleviate this burden, and also to improve administrator–research

worker interactions, students and faculty should be involved in both monitoring and establishing safety procedures, perhaps by the appointment of one or more faculty members as "safety advocates" rather than safety officers, and by membership on a departmental safety committee.

5. It is essential that some kind of laboratory inspection schedule—without prior announcement—be established. If these are handled in a collegial way, the inspections can have a positive effect on the development of laboratory safety culture. The inspections, as well as other interactions with departmental safety committees and/or advocates, could also play a role in encouraging intra-lab coaching/collaboration and teaching researchers how to politely approach their peers about potential safety hazards that should be corrected. In this way, a positive safety learning environment in the laboratory can be created.

INFLUENCES FROM THE BOTTOM UP

Currently most departments require formal safety training for incoming graduate students. This typically involves communicating information on the proper use of protective gear, such as lab coats, safety glasses, proper foot and head protection, and fume hoods, along with scenarios of accidents that have occurred when such precautions were not taken. It also provides information about what to do when an accident occurs, which requires knowledge of emergency phone numbers, location of safety showers, etc., and includes hands-on training in the use of fire extinguishers. Some of this training includes advice about what to do in case of a fire or the occurrence of a natural disaster. It is important to include instructions about procedures to follow in the event of chemical spills and explosions.

For institutions that are still not providing such training, it should be made part of the curriculum. In addition, it should be a requirement not only for students, but for postdocs and other researchers as well. Training for non-student researchers hired directly by the principal investigator that may not arrive on a specific schedule, as graduate students usually do, and may not pass through an institutional safety training program is a concern. Particularly problematical are research workers who enter university laboratories with their own funding, which often means that it may not be possible to use a payroll roster to screen them for safety training. However, in nearly all cases that we are aware of, research workers are given keys or electronic card access to the buildings and laboratories in which they work. The key-issuing office may be used as a checkpoint for determining whether incoming laboratory workers have received appropriate safety training, irrespective of whether they claim to have had such training in another institution. They should not be given their building access until someone has signed off on this training.

Recalcitrant Group Leaders and/or Co-workers

While this report is written to help improve the safety culture of academic laboratories in the United States, it is realistic to recognize that there will be a minority of principal investigators and research groups that will be resistant to the development of a positive safety culture in their laboratories. In such situations, the ultimate responsibility for ensuring safe working conditions rests with the department chair and the university administration. Although the prospect of shutting down a principal investigator's laboratory is an unfortunate action, it cannot be taken completely off the table as a last resort way of making sure that research workers in a university are protected.

IDEAS TO ADDRESS SAFETY DYNAMICS

Advantages for Recruiting and Laboratory Funding

There are many advantages in promoting a safety culture and environment in a chemistry department. Ultimately, the results of providing a culture of safe and reliable scientific practices can be leveraged to enhance the overall success of a chemistry department and possibly increase its competitiveness. While there are some numerical metrics that might be used to characterize the success of a particular department, one measure might be the quality and competitiveness of the department to attract the very best talent in chemistry. For example, if a particular department is noted for establishing a good and safe culture in doing scientific research, then this may attract highly competitive graduate students through their recruiting efforts. Most chemistry departments have extensive events to recruit excellent students each year to their programs. Some of these events are extremely costly to these departments. The department could optimize their investments in this process by including information in the message to future students that ONLY a safe and welcoming culture in doing scientific research would be allowed in their department. This message, and the data to support the claim, would be a very powerful point to make to those intending to do research as they undertake careers in the sciences and would offer respect and assurance to the students that the department cares about their safety. Often, the issues of laboratory safety are overlooked during recruiting weekends and events, and providing this information to prospective students (and postdocs) would make a strong point that their safety is important to that particular department. This approach may also be an advantage for future grant proposals by the department, both external (federal) and internal (university-wide).

Safety in Departmental Rankings

While it is clear that many departments gain more resources by virtue of their accomplishments in publishing papers, acquiring research grants, and ultimately in national rankings, there should be a level of appreciation and reward for practicing safe methods in doing research as well. It is the responsibility of the entire university community to promote safe environments for research. A measure of its impact would be to have safety as a measure of success of a department or college. Because resources are heavily contended throughout the university, the administration or leadership could set a standard that it expects its faculty to uphold in providing a safe environment for the many students that do research. As mentioned previously, the faculty salary program could have safety as a measure of success. Also, the resources used for start-up funds and other renovations could be allotted in part based on a department's or unit's safety practices.

Role of the Principal Investigator

In regard to creating a culture that is conducive to safety, there needs to be a nonthreatening atmosphere. This requires the principal investigator to be able to make observations and, subsequently, suggestions in a proactive manner. If the methods mentioned above regarding near-miss reporting are to make an impact, then it should *not* matter who reported the incident or who is the primary person involved. Instead, the focus should be on addressing the threat of the danger and eliminating it as quickly as possible. If a culture that is created for doing research safely is to be successful, this step is critical. It makes all those involved know that everyone is responsible and that no one should harm themselves in the important research they are doing.

SKILLS AND TOOLS FOR PRINCIPAL INVESTIGATORS

Competitive academic programs of teaching and research require investments not only in buildings, equipment, and infrastructure, but also in excellent personnel (senior leaders, faculty, staff). These personnel need both scientific expertise and skills of leadership and management to establish and sustain a strong, positive safety culture. Provosts, deans, and chairs need to work with faculty who lead academic research laboratories

BOX 4-2
University of Minnesota Safety Program

One approach to changing academic laboratory safety culture is illustrated by the collaboration among the Department of Chemistry and the Department of Chemical Engineering & Materials Science at the University of Minnesota (Twin Cities, MN) and the Dow Chemical Company (Midland, MI).[a] Faculty, department chairs, graduate students, and postdoctoral associates from the departments partnered with EHS professionals to develop awareness and practices to foster safety. This "bottom-up" approach was developed and implemented by groups of volunteer laboratory safety officers (LSOs)—graduate students and postdoctoral associates—from each of the 59 research laboratories housed in the two departments. Organized as a "joint safety team" and charged with developing the program, the LSOs developed a safety approach focused on day-to-day attitudes, values, and practices.

Initial activities of the LSOs included surveys of safety attitudes and practices among faculty, staff, and students; tours of a wide variety of other laboratories in their home institution, as well as a visit to the Dow facilities in Midland. Each of these activities created opportunities for the LSOs to align their perceptions and expectations about safety practices with actual laboratory practices that were both inside and outside of their own areas of specialization, and in a research facility outside an academic department.

Supported by seed funding from the heads of the two departments, the LSOs developed a set of recommendations (CARE) focused on four areas: Compliance (roles and expectations), Awareness (signage, regular discussion of safety, e-mail updates), Resources (equipment, infrastructure, waste management), and Education (with a particular focus on lab-specific topics). Initial activities, again formulated by the LSOs, targeted areas such as peer tours of laboratories, personal protective equipment, a public website (http://www.jst.umn.edu/), and a lab cleanup week. The LSOs also instituted a practice of beginning group meetings and all departmental seminars with "safety moments" and, as an example, created an illustrative slide (Figure 4-1) that contains a safety topic relevant to the group or seminar topic, educational content, and one or more key citations. These "safety moments" are a striking example of strategies that make safety topics normal parts of academic culture and direct attention to the practice and science of safety.

The UMN program is exciting, but it remains to be seen how it will affect the overall safety of the department over time. More data are needed about whether

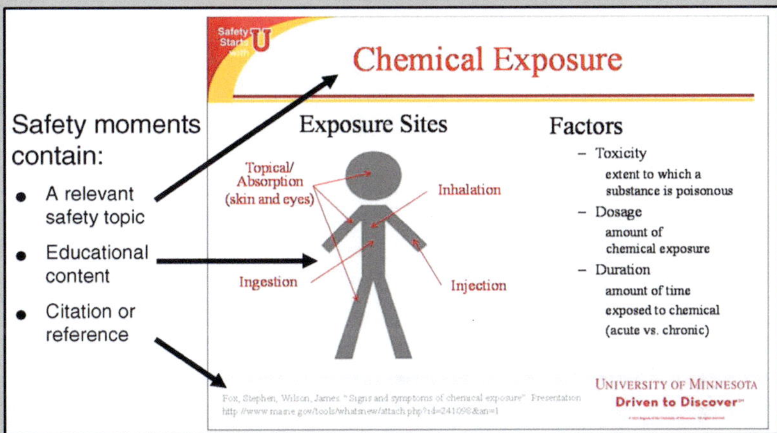

FIGURE 4-1 An example slide that may be developed for discussion around a safety topic.
SOURCE: Reprinted with permission from McGarry, K.A, et. al. Student Involvement in Improving the Culture of Safety in Academic Laboratories. J. Chem. Educ. 2013, 90(11): 1414-1417. Copyright 2013. American Chemical Society.

it is sustainable and scalable within its home institution, whether it will produce long-term changes in its home institution, or which of its features will be adaptable or critical to other departments and institutions. McGarry et al.[b] suggest several features that may be important to the program's success and sustainability, but emphasize thatthe program's characteristic "bottom-up" approach may be particularly important as it builds on the drive and future focus of the next generation of academic scientists.

---

[a]University of Minnesota, Department of Chemistry. "Dow + U = lab safety." Available at http://discover.umn.edu/news/vision-leadership/u-and-dow-chemical-team-lab-safety/. Accessed March 12, 2014. [b]McGarry, K.A, et. al. Student Involvement in Improving the Culture of Safety in Academic Laboratories. *J. Chem. Educ.* 2013, 90(11): 1414-1417.

to identify the variety of leadership challenges they face and provide explicit tools and professional development opportunities that address these challenges. Useful tools and professional development include

- Resources for hazard analysis,[22,23] which might include support for faculty to attend workshops on hazard analysis offered by groups such as the ACS Division of Chemical Health and Safety and expectations that they do so as part of their faculty role.
- Introductions to guidance and processes available from institutional human resources and mental health services, with examples of how to approach difficult personnel issues and of when and how individuals can be referred to the services.
- Development and mentorship programs focused on leadership skills. Resources such as *Making the Right Moves*,[24] *At the Helm*,[25] and the American Association for the Advancement of Science Careers site[26] contain practical information and guidance from experienced researchers; resources such as *Training Scientists to Make the Right Moves*[27] provides guidance for institutions, and institutional programs can be tailored to specific challenges faced by faculty in a local environment.
- Institutional support for development and dissemination of lab-specific safety information, for expectations that faculty and trainees will regularly include EHS professionals in research planning, and for involvement of students and postdoctoral students in safety programs (indeed, the Minnesota program described in Box 4-2 suggests that institutions may need to empower and support trainees as leaders of departmental programs).

---

[22] National Research Council. *Prudent Practices in the Laboratory: Handling and Management of Chemical Hazards, Updated Version*. The National Academies Press, Washington, DC, 2011.

[23] American Chemical Society Committee on Chemical Safety. *Identifying and Evaluating Hazards in Research Laboratories: Guidelines Developed by the Hazards Identification and Evaluation Task Force*. American Chemical Society, Washington, DC, 2013.

[24] Burroughs Wellcome Fund and Howard Hughes Medical Institute. *Making the Right Moves: A Practical Guide to Scientific Management for Postdocs and New Faculty,* 2nd Ed. Available at http://www.hhmi.org/sites/default/files/Educational%20Materials/Lab%20Management/Making%20the%20Right%20Moves/moves2.pdf.

[25] Barker, K. *At the Helm: Leading Your Laboratory*, 2nd Ed. Cold Spring Harbor Laboratory Press, Long Island, NY, 2010.

[26] Bea, R., I. Mitroff, D. Farber, Howard Foster, and K. H. Roberts. A new approach to risk: The implications of E3. *Risk Management* 2009; 11(1): 30-43.

[27] Burroughs Wellcome Fund and Howard Hughes Medical Institute. *Training Scientists to Make the Right Moves: A Practical Guide to Developing Programs in Scientific Management.* 2006. Available at http://www.hhmi.org/sites/default/files/Educational%20Materials/Lab%20Management/Training%20Scientists/training-scientists-fulltext.pdf.

- Integration of safety work into promotion and recognition programs at all levels of the institution, so that the work required to advance academic laboratory safety becomes a "normal" part of performance expectations and of academic discourse (group meetings, seminars, dissertations, publications).

5

Findings, Conclusions, and Recommendations

BEYOND ACADEMIC CHEMISTRY LABORATORIES

The statement of task for this study sets clear boundaries regarding academic chemistry research laboratories. However, it is worth noting that many of the same risks and hazards identified in this report exist under the same cultural constraints in other research communities within colleges and universities. Moreover, both research and non-research laboratories in non-academic settings may carry similar risks and constraints. Application of the analyses and changes suggested herein may be helpful in these other settings as well.

Researchers Beyond Chemistry Research

Clearly other research units in colleges and universities are affected by the organizational factors outlined in this report. Organizational structure, reporting relationships, evaluation criteria, funding and time pressures, workload, and workplace stress are not unique to chemistry research. It is paramount to safeguard the welfare of the students, staff, and faculty and to establish expectations and support systems that enable them to work safely. While the specific hazards of different research units may vary, the organizational and system processes remain the same. Therefore, many of these recommendations can be generalized to other research units within the academic sector.

Beyond Academic Laboratories

While many industrial and non-academic research laboratories provide excellent examples of safety culture, it is also true that there are many that can benefit from these recommendations. The system processes that govern safety culture operate across contexts, and the need for careful consideration of whether institutional practices support safety is independent of the university/non-university context. Designing institutional systems so that they promote the ability of all individuals to take the actions needed to work safely is critical to the twin goals of promoting the nation's scientific stature and the health and safety of the people who produce it.

If viewed as a system, these recommendations for improving the culture of safety can be applied broadly and can allow the greater community to solve problems while simultaneously advancing productivity, safety, and sustainability across a wide range of settings.

FOCUS ON CHEMICAL RESEARCH: FINDINGS, CONCLUSIONS, AND RECOMMENDATIONS

In response to the statement of task and building on the discussion in the preceding chapters, a series of findings have been identified, conclusions made, and recommendations presented. They are presented under four categorical headings: *Institution-wide Dynamics and Resources; Research Group Dynamics; Data, Hazard Identification, and Analysis;* and *Training and Learning.*

Institution-Wide Dynamics and Resources

The broad institutional setting in which research takes place can strongly influence whether university laboratories develop and sustain a strong, positive safety culture. Specifically, the level of importance attached to safety by university leadership, the way these leaders promote safety as a core institutional value, the way they direct resources, and the structure of incentives and reporting relationships they support all affect the degree of priority given to safety practices. The list of findings, conclusions, and recommendations below address issues of *Institution-Wide Dynamics and Resources.*

> **Finding 1:** *Safety is emerging as a priority and a core value of many academic institutions and of individual laboratories. A strong, positive safety culture is more beneficial than a compliance-only culture.*

Finding 2: *A strong, positive safety culture is a core element in the responsible conduct of research.*

Conclusion 1: *If laboratory safety is an unquestioned core value and operational priority for the institution, then safety will never be traded for research productivity.*

Recommendation 1: The president and other institutional leaders must actively demonstrate that safety is a core value of the institution and show an ongoing commitment to it.

Finding 3: *The availability and commitment of university resources to laboratory safety vary across institutions.*

Finding 4: *Universities often do not provide sufficient incentives to promote a strong, positive safety culture. In some cases they may create barriers or disincentives.*

Conclusion 2: *University policies and resource allocations have a strong impact on a department's ability and willingness to help provide for a strong, positive safety culture. If an institution or individual laboratory wants to develop and sustain a safe and successful research program, then it must consider the resources it has available for safety and explore research options and requirements accordingly.*

Recommendation 2: The provost or chief academic officer, in collaboration with faculty governance, should incorporate fostering a strong, positive safety culture as an element in the criteria for promotion, tenure, and salary decisions for faculty.

Recommendation 3: All institutions face a challenge of limited resources. Within this constraint, institutional head(s) of research and department chairs should consider the resources they have available for safety when considering or designing programs, and identify types of research that can be done safely with available and projected resources and infrastructure.

Finding 5: *There is a lack of clarity and consistency about safety roles and responsibilities across the university, particularly among faculty, researchers, and environmental health and safety personnel.*

Recommendation 4: University presidents and chancellors should establish policy and deploy resources to maximize a

strong, positive safety culture. Each institution should have a comprehensive risk management plan for laboratory safety that addresses prevention, mitigation, and emergency response. These leaders should develop risk management plans and mechanisms with input from faculty, students, environmental health and safety staff, and administrative stakeholders and ensure that other university leaders, including provosts, vice presidents for research, deans, chief administrative officers, and department chairs, do so as well.

Research Group Dynamics

Many research groups have differential power dynamics, which, if not appropriately addressed, can work against the development of a strong, positive safety culture. Department chairs and principal investigators should take steps to change these dynamics, creating mechanisms that empower laboratory researchers to communicate freely about safety and take an active role in establishing and promoting a strong, positive safety culture and in sustaining a safe research enterprise. The list of findings, conclusions, and recommendations below address issues of *Research Group Dynamics*.

Finding 6: *There is variability across academia with regard to the involvement of researchers at all levels in establishing and sustaining a strong, positive laboratory safety culture.*

Finding 7: *The deeply rooted hierarchy and highly competitive nature of academic research can inhibit the advancement of a strong, positive safety culture.*

Finding 8: *Students and postdocs are dependent on the principal investigator for their professional advancement. The power differential in this relationship may affect group members' willingness to raise safety concerns.*

Finding 9: *Most researchers in academia are still in the early phases of their professional development. As such, they may not have the requisite knowledge and skills to recognize and understand the risks associated with their work.*

Finding 10: *Research is regularly performed independently (including during off-hours and alone) and may be carried out with limited or no oversight or feedback.*

Conclusion 3: *Contribution and engagement by both principal investigators and by researchers through an open and ongoing dialogue are critical to creating a strong, positive safety culture. Safety culture is more likely to be sustained when safety issues are discussed broadly and frequently as an integral part of the research training and development process.*

Conclusion 4: *There are several key attributes related to research group dynamics that contribute to the advancement of the laboratory safety culture. A strong, positive safety culture*

- *includes open communication about safety as a key element that is sought out, valued, and acted upon;*
- *values learning and continuous improvement with respect to safety;*
- *includes regular safety communication, for example, "safety moments," in academic research events (e.g., seminars, group meetings, doctoral defenses, and teaching); and*
- *empowers student and research trainees to have a "voice" and maintain an environment that encourages raising safety concerns freely without fear of repercussions.*

Conclusion 5: *A research group with a strong, positive safety culture engages with environmental health and safety personnel collaboratively.*

Recommendation 5: Department chairs and principal investigators should make greater use of teams, groups, and other engagement strategies and institutional support organizations (e.g., environmental health and safety, facilities), to establish and promote a strong, positive, safety culture.

Recommendation 6: Department chairs should provide a mechanism for creating a robust safety collaboration between researchers, principal investigators, and environmental health and safety personnel.

Data, Hazard Identification, and Analysis

In addition to improving the organizational dynamics that drive safety practice, laboratories have a need for data and to conduct analyses that will help them identify and mitigate hazards. Traditionally, safety performance has been measured using lagging or after-the-fact indicators, such as numbers of accidents and lost-time injuries. To change behavior and culture before an incident occurs, organizations may take advantage of *leading indicators*: before-the-fact data that can help identify risks and vulnerabilities ahead of time. One key approach to identify hazards before they cause harm is to report and collect data on near misses. Another way to identify hazards is to conduct hazard analysis, a process to assess risks and their consequences and ensure that they are mitigated or eliminated before any lab work is initiated. The list of findings, conclusions, and recommendations below address issues of *Data, Hazard Identification, and Analysis*.

Finding 11: *Leading indicators from hazard analysis, risk mitigation, and best practices are not being widely used in laboratory safety planning. Often these data are not being collected for academic and non-industrial laboratories.*

Finding 12: *Incident and near-miss data are important sources of information for driving improved safety performance and for monitoring progress. Such key data are often repressed or distorted when there is a punitive approach in response to incidents.*

Conclusion 6: *Information is a key input to establishing and promoting a strong, positive safety culture. Incident and near-miss reports are important learning tools for laboratory safety, but presently are not effectively reported, compiled, analyzed, and disseminated within the research community. To ensure that useful data are available, a change in reporting and the availability and sharing of information is necessary.*

Recommendation 7: Organizations should incorporate non-punitive incident and near-miss reporting as part of their safety cultures. The American Chemical Society, Association of American Universities, Association of Public and Land-grant Universities, and American Council on Education should work together to establish and maintain an anonymous reporting system, building on industry efforts, for centralizing the collection of information about and lessons learned from incidents and near misses in academic laboratories, and linking these data to the scientific literature. Department chairs and university leadership should incorporate the use of this system into their safety planning. Principal investigators should require their students to utilize this system.

Finding 13: *Researchers may not understand or appreciate the hazards of chemical materials and procedures in their work. This may be* *especially relevant for departments in which researchers typically have less training in chemistry (e.g., molecular biology, biochemistry, and engineering), yet often work with potentially hazardous materials or procedures.*

Finding 14: *Hazard analysis is not routinely incorporated into experimental designs, procedures, and records in academia.*

Conclusion 7: *Routine hazard analysis is a critical component in research planning and execution. It represents an element of a strong, positive safety culture. Comprehensive hazard analysis and the use of engineering controls are especially important for experiments that are new to the individual and/or are being scaled up.*

Recommendation 8: The researcher and principal investigator should incorporate hazard analysis into laboratory notebooks prior to experiments, integrate hazard analysis into the research process, and ensure that it is specific to the laboratory and research topic area.

Training and Learning

Training in safety practices—both initial training and ongoing mentoring and support—is an essential element in developing and sustaining a strong, positive safety culture. This is particularly important with researchers in academic labs, who are often relatively young and have limited experience. Entering (and even experienced) students may not know how to assess the risks of what they are doing, how to assess changes in risks if they change a key experimental parameter, or how to keep a small error from causing major problems. Moreover, they may not realize that a process they used in the past without apparent incident was out of the ordinary or dangerous. The list of findings, conclusions, and recommendations below address issues of *Training and Learning*.

Finding 15: *Laboratory safety training is highly variable across institutions, departments, and research groups.*

Conclusion 8: *A high-quality training program is an important element of a strong, positive safety culture.*

Finding 16: *There is a lack of comprehensive, early, and individual-laboratory-centric training and education for researchers, principal investigators, and in some cases, environmental health and safety staff. Many* *researchers arrive at a new institution or in a new laboratory without proper training or appreciation for appropriate safe laboratory practice.*

Conclusion 9: *Classroom and online training is necessary but not sufficient to ensure knowledge, skills, qualifications, and abilities to perform safely in a laboratory environment and to establish a strong, positive safety culture.*

Recommendation 9: Department leaders and principal investigators, in partnership with environmental health and safety personnel, should develop and implement actions and activities to complement initial, ongoing, and periodic refresher training. This training should ensure understanding and the ability to execute proper protective measures to mitigate potential hazards and associated risks.

Appendix

Biographies of Committee Members and Staff

Holden Thorp obtained his B.S. in chemistry from UNC-Chapel Hill in 1986, his Ph.D. in chemistry from Caltech in 1989, and was a postdoctoral associate at Yale University. He came back to UNC-Chapel Hill as assistant professor in 1993. In July 2008, he became the 10th chancellor of UNC-Chapel Hill. In 2013, he became the provost and distinguished professor of chemistry and medicine at Washington University in St. Louis. Dr. Thorp is on the National Security Higher Education Advisory Board, the Board of Directors of Barnes-Jewish Hospital, and the Board of Trustees of the National Humanities Center. Thorp co-authored "Engines of Innovation—The Entrepreneurial University in the 21st Century," a UNC Press book that makes the case for the pivotal role of research universities as agents of societal change. He has published 130 scholarly articles on the electronic properties of DNA and RNA, holds 12 issued U.S. patents and co-founded Viamet Pharmaceuticals, which is developing drugs for prostate cancer and fungal infections.

David DeJoy is Professor Emeritus of Health Promotion and Behavior and Director Emeritus of the Workplace Health Group in the College of Public Health at the University of Georgia. Dr. DeJoy has over 30 years of experience in workplace safety and health as a researcher, instructor, and consultant. His areas of research include safety climate/culture, work organization, safe work practices, risk communication, and theory-based intervention design/intervention effectiveness. He has published approximately 120 scientific articles and book chapters and he has presented over

200 papers at scientific and professional meetings. Editorial board service includes *Safety Science*, the *Journal of Safety Research*, the *Journal of Occupational Health Psychology*, and the National Safety Council Press. Honors include the Liberty Mutual Prize for research in occupational safety and ergonomics, the Liberty Mutual Medal for research in occupational safety and ergonomics, and the Williams A Owens Award for research in the social-behavioral sciences. Extramural funding for his research has come from CDC, FEMA, NIH, and NIOSH. Dr. DeJoy has served on numerous expert panels, review committees, and advisory panels at the national and international levels. He holds a Ph.D. from the Pennsylvania State University.

John Bercaw received his B.S. degree from North Carolina State University in 1967, his Ph.D. from the University of Michigan in 1971, and undertook postdoctoral research at the University of Chicago. He joined the faculty at the California Institute of Technology as an Arthur Amos Noyes Research Fellow in 1972, and in 1974 he joined the professorial ranks, becoming professor of chemistry in 1979. From 1985 to 1990 he was the Shell Distinguished Professor of Chemistry, and in 1993 he was named Centennial Professor of Chemistry. Bercaw has been a Seaborg Scholar at Los Alamos National Laboratory (2004), the Robert Burns Woodward Visiting Professor at Harvard University (1999), The George F. Baker Lecturer at Cornell University (1993), Visiting Miller Professor at the University of California, Berkeley (1990), and a Royal Society of Chemistry Guest Research Fellow at Oxford University (1989-1990). From 2009 to 2012 he was also KFUPM Visiting Chair Professor at King Fahd University of Petroleum and Minerals. He has served on numerous panels for the Department of Energy and the National Research Council, and beginning in 1999 has been a member of the Science and Technology Committees for national laboratories: Los Alamos National Security and Lawrence Livermore National Security. Bercaw is a Fellow of the American Association for the Advancement of Science (1986), a member of the National Academy of Sciences (1990), a Fellow of the American Academy of Arts and Sciences (1991), and was awarded an Honorary Doctorate of Science from the University of Chicago in 2001. His research interests are in synthetic, structural, and mechanistic organotransition metal chemistry. Investigations include catalysts for polymerization and selective trimerization of olefins, investigations of hydrocarbon partial oxidation with transition metal complexes, and the development of catalysts for syngas and light alkane conversions to chemicals and fuels. He has published approximately 300 peer-reviewed scientific articles.

Robert Bergman completed his undergraduate studies in chemistry at Carleton College in 1963 and received his Ph.D. at the University of Wisconsin in 1966 under the direction of Jerome A. Berson. Bergman spent 1966-1967 as a North Atlantic Treaty Organization Fellow in Ronald Breslow's laboratories at Columbia, and following that began his independent career at the California Institute of Technology. He accepted an appointment as professor of chemistry at the University of California, Berkeley, in July 1977, and moved his research group to Berkeley about a year later. In 2002 he was appointed Gerald E. K. Branch Distinguished Professor. He has received a number of national awards and has coauthored more than 500 publications in peer-reviewed journals. Bergman was trained as an organic chemist and spent the first part of his independent career studying reaction mechanisms that involve unusually reactive molecules, such as 1,3-diradicals and vinyl cations. In 1972 he discovered a transformation of ene-diynes that was later identified as a crucial DNA-cleaving reaction in several antibiotics that bind to nucleic acids. In the mid-1970s Bergman's research broadened to include organometallic chemistry, which led to contributions to the development and study of the reaction mechanisms of migratory insertion and oxidative addition reactions, the chemistry of new dinuclear complexes, and the investigation of organometallic compounds having metal-oxygen and -nitrogen bonds. He is probably best known for his discovery of the first soluble organometallic complexes that undergo intermolecular insertion of transition metals into the carbon-hydrogen bonds of alkanes. Most recently he has been involved in collaborative studies with colleagues at Berkeley and elsewhere that include applications of catalytic C-H activation reactions in organic synthesis, reactions catalyzed by supramolecular systems, the chemistry of complexes bearing metal-heteroatom single and multiple bonds, and methods for the conversion of polyhydroxy compounds into materials currently derived from petroleum.

Joseph Deeb holds a Ph.D. in industrial engineering specializing in human factors and ergonomics. Dr. Deeb is a Certified Professional Ergonomist (CPE) and a Registered Member of the Ergonomics Society of the UK (M.Erg.S.). He has over 27 years of both academic and industry experience. He has been with ExxonMobil for over 21 years. His role is the Human Factors Advisor and Lead in the ExxonMobil Human Factors Center of Excellence. The Human Factors Center of Excellence (HFCOE) provides leadership in the effective use of human factors and ergonomics to achieve outstanding operational performance. The HFCOE proactively identifies risks and associated control practices across business functions and operations. Additionally, Dr. Deeb has expertise in risk perception and risk tolerance areas and their applications and techniques in the development of systems and guidance to improve safety performance. These applications and techniques engage individuals to identify, evaluate, and execute safe behavior and, to approach others during a safe or unsafe behavior, to provide constructive input and coaching.

Larry Gibbs is Associate Vice-Provost for Environmental Health and Safety at Stanford University where he is responsible for health, safety, and environmental risk management programs as well as oversight of institutional emergency planning and risk communication. In addition to a campus population of over 10,000 employees and 17,000 undergraduate and professional students, Stanford has 2500 laboratories involved annually in over $700 million of research ranging from basic sciences and engineering to medical and human subjects clinical research. His responsibilities include overall campus health and safety management and oversight of hazardous chemical, radiological, and biological materials and physical agents used in research and throughout Stanford. Mr. Gibbs is a lecturer at the Woods Institute for the Environment at Stanford and serves on the Stanford Board of Overseers for the SLAC National Accelerator Laboratory. He has graduate degrees in science education from Boston University and in industrial hygiene and public health from the University of Michigan. Mr. Gibbs is a certified industrial hygienist with over 25 years of experience in academic, research, and clinical institutions. In addition to his work at the university, he has served as a consultant for industrial, pharmaceutical, biotechnology, and government organizations and currently serves on the scientific advisory board for nanoTox, Inc., a nanomaterials safety, testing, and consulting firm. He has authored over 25 publications, co-authored two books, served as officer and board member in a number of national and international professional associations, including Chair of the ACGIH in 2008. Larry is a Fellow of the American Industrial Hygiene Association. He recently chaired the statewide California Higher Education–DTSC–NIOSH Working Group that developed and published the *NanoToolkit: Working Safely with Engineered Nanomaterials in Academic Research Settings*.

Theodore Goodson III received his B.A. in 1991 from Wabash College and earned his Ph.D. in chemistry at the University of Nebraska-Lincoln in 1996. After postdoctoral positions at the University of Chicago and at the University of Oxford, he accepted a position as assistant professor of chemistry at Wayne State University in 1998. In 2004, he moved to the University of Michigan as professor of chemistry. In 2008, he was appointed as the Richard Barry Bernstein Professor of Chemistry at the University of Michigan. Dr. Goodson's research centers on the investigation of nonlinear optical and energy transfer in organic multi-chromophore systems for particular optical and electronic applications. His research has been translated into technology in the areas of two-photon organic materials for eye and sensor protection, large dielectric and energy storage effects in organic macromolecular materials, and the detection of energetic (explosive) devices by nonlinear optical methods. He has investigated new quantum optical effects in organic systems which have applications in discrete communication systems and sensing. In 2009, he founded Wolverine Energy Solutions and Technology Inc., a start-up company with contracts to produce high energy density capacitors for military, automotive, and medical devices. He has also developed and translated a new system for the detection of IEDs remotely. Some of Dr. Goodson's awards include the Distinguished Faculty Achievement Award, the National Science Foundation American Innovation Fellowship, National Science Foundation CAREER Award, Alfred P. Sloan Research Fellowship, Camille Dreyfus Teacher-Scholar Award, Lloyd Ferguson Young Scientist Award, The Percy Julian Award, American Chemical Society Fellow, The American Association for the Advancement of Science Fellow, Imes and Moore Mentorship Award, American Chemical Society Minority Mentorship Award, University Faculty Recognition Award, College of Science Teaching Award, and a National Academy of Sciences Ford Postdoctoral Fellowship. Dr. Goodson has been a Senior Editor for the *Journal of Physical Chemistry* since 2007.

Andrew Imada specializes in human and organizational change. He works with people and organizations to change their safety cultures, respond to scalability demands, implement enterprise resource planning systems, and survive generational transitions. He teaches them to achieve these successes by balancing productivity, safety, quality, and human needs. Dr. Imada has provided consulting services to a wide range of clients including AT&T, Aramark, British Columbia Telephone, Chevron Americas Products Company, Chevron Production Company, Hamersley Iron, Iron Mountain, Los Angeles Dodgers, NASA, PG&E, Sheraton Hotels, Pacific Coast Building Products, Sierra Nevada Brewing, Southern Wine and Spirits, Teichert Inc., U.S. Army, and Lawrence Berkeley National Laboratories. He served as Senior Scientific Advisor for the Steelcase User Center Design Group and worked on projects advising the National Research Council, International Labour Office, and the University of California. He is a Certified Professional Ergonomist. From 2009 to 2012 he served as the President of the International Ergonomics Association, which represents 49 federated societies and more than 25,000 ergonomists. He will be the President for the Human Factors and Ergonomics Society in 2014-2015. Dr. Imada won the 1998 Liberty Mutual

Prize and the 2000 Liberty Mutual Medal in international competitions for occupational safety and ergonomics research. Dr. Imada was a professor of ergonomics and safety sciences at the University of Southern California for 19 years. He also served as the Director of the USC Safety Science Center and the International Distance Learning Liaison at the USC Center for Scholarly Technology. He has published extensively and edited a book entitled, *Participatory Ergonomics*. He was a visiting scholar at Luleå University in Sweden and taught graduate courses on participatory strategies for improving safety, ergonomics, and productivity. Dr. Imada serves on the National Research Council's Board on Human Systems Integration (BOHSI). He served on the Board of Consulting Editors for the *Journal of Applied Psychology* and is a technical reviewer for professional journals. He served as a director on the Board of Certification in Professional Ergonomics. He is a Fellow of the Human Factors and Ergonomics Society and the International Ergonomics Association. Dr. Imada received a Rotary Foundation International Fellowship to conduct research at the University of Sussex in England. He earned his bachelor of arts in psychology and minored in business from the University of San Francisco and his masters and doctoral degrees in industrial and organizational psychology from The Ohio State University.

Kimberly Jeskie is the Directorate Operations Manager for Facilities and Operations and the Director of the Integrated Operations Support Division for the Oak Ridge National Laboratory (ORNL). She has 23 years of experience at ORNL, beginning her career as a research technician in physical organic chemistry. Over the years, she has held several roles within the areas of environmental protection, waste management, radiological control, facility management, performance assessment, training, and safety all in direct support to the research community. Ms. Jeskie has been trained in the principles of accident investigation and human performance fundamentals and has participated in and led a number of event investigations within ORNL and at other Department of Energy facilities. In her current role, she is responsible for the work planning and hazards analysis systems and tools utilized by both principal investigators and operations personnel across ORNL. The Integrated Operations Support Division, which she directs, also provides the systems, tools, and performance analysis for ensuring integrated facilities management at the Laboratory. Ms. Jeskie holds a bachelor of science in chemistry and mathematics from Cumberland College and a masters in public health with an emphasis in occupational safety and health management from Tulane University. She is a Past Chair of the ACS Division of Chemical Health and Safety and an Associate with the ACS Committee on Chemical Safety, heading the task force creating guidance on hazards analysis techniques that can be applied in the research environment.

Bradley Pentelute joined the MIT Chemistry faculty in July 2011 after a 3-year postdoctoral appointment in the group of Professor R. John Collier at Harvard Medical School. He obtained his Ph.D. degree in organic chemistry in 2008 under the guidance of Professor Stephen Kent at the University of Chicago. He is currently the Pfizer-Laubach Career Development Research Professor at MIT and is also an associate member of the Broad Institute. The Pentelute lab develops new technologies to deliver polypeptides and proteins into cells by the use of bacterial agents including Anthrax toxin. The lab also develops new chemical technologies for the macrocyclization of peptides. Lastly, the lab uses fast flow chemical methods to synthesize and study mirror image proteins.

Karlene Roberts is a professor at the Walter A. Haas School of Business, at the University of California at Berkeley and Director of the Center for Catastrophic Risk Management at Berkeley. Roberts earned her bachelor's degree in psychology from Stanford University and her Ph.D. in Industrial psychology from the University of California at Berkeley. She also received the docteur honoris causa from the Universite Paul Cezanne (Aix Marseilles III). Since 1984 Roberts has investigated the design and management of organizations and systems of organizations in which error can result in catastrophic consequences. She has studied both organizations that failed and those that succeed in this category. Some of the industries Roberts has worked in are the military, commercial marine transportation, health care, railroads, petroleum production, commercial aviation, banking, and community emergency services.

Jennifer Schomaker is currently an assistant professor at the University of Wisconsin-Madison, where she began her independent career in 2009. She received her bachelor's degree in chemistry from Saginaw Valley State University while she was employed at the Dow Chemical Company in Midland, Michigan. Her early research at Dow in the Organic Chemicals and Polymer Laboratory involved the development of biocatalytic methods for the synthesis of enantiomerically pure monomers. She then moved to the Agricultural Chemicals Process Research group where she participated in the route selection and scale-up campaigns for two new herbicides. After leaving Dow Chemical, she began her doctoral studies at Michigan State University in the laboratory of Professor Babak Borhan, focusing on new methodologies for the preparation of heterocycles, as well as the total syntheses of the haterumalides. After completing her Ph.D. in 2006, she moved to Berkeley as an NIH postdoctoral fellow in the labs of Professor Robert G. Bergman, collaborating with Professor F. Dean Toste on the development of cobalt dinitrosoalkane complexes to enable the mild functionalization of the C-H bonds of alkenes. Her work at UW-Madison is centered on the development of new methods for the mild functionalization of hydrocarbons using first-row and coinage metal catalysts.

Alice Young is Associate Vice President for Research and professor of psychological sciences at Texas Tech University (TTU) and of Pharmacology and Neuroscience at Texas Tech University Health Sciences Center. As Associate Vice President for Research, she works with TTU responsible research committees and the TTU Office of Environmental Health and Safety. Before joining the Texas Tech University System in 2004, Dr. Young was professor of psychology and of psychiatry and behavioral neurosciences at Wayne State University, where she served as Associate Dean for Research and Graduate Programs in the College of Science from 1996 to 2004. Her research and teaching focus on behavioral and brain processes involved in the actions of psychoactive drugs, with over 20 years of NIH support for studies of drug tolerance and dependence. Her professional service has included service as Associate Editor of *Behavioural Pharmacology* and *The Journal of Pharmacology and Experimental Therapeutics*, as a member of ADAMHA and NIH review panels, and as a member of the Board of Scientific Affairs of the American Psychological Association and the Board of Directors of the College on Problems of Drug Dependence. Dr. Young earned a doctorate in experimental psychology from the University of Minnesota in Minneapolis and received postdoctoral training in pharmacology at the University of Michigan in Ann Arbor.

Staff

Douglas Friedman is a senior program officer with the Board on Chemical Sciences and Technology at the National Research Council (NRC) of the National Academy of Sciences. His primary scientific interests lie in the fields of organic and bio-organic materials and chemical and biological sensing and nanotechnology, particularly as they apply to national and homeland security. Dr. Friedman has supported a diverse array of activities since joining the NRC. He has directed studies in the areas of carbohydrate chemistry and glycobiology, crude oil pipeline transportation, computational molecular dynamics simulations, chemical and biological defense, and technological surprise. Dr. Friedman has also supported activities in biomass utilization, critical resources, and antibiotics research and development. Prior to joining the NRC, Dr. Friedman performed research in physical organic chemistry and chemical biology at Northwestern University, the University of California, Los Angeles, the University of California, Berkeley, and Solulink Biosciences. He holds a Ph.D. in chemistry from Northwestern University and a bachelor's degree in chemical biology from the University of California, Berkeley.

Toby Warden has a Ph.D. in social ecology with an emphasis on environmental analysis and design from the University of California, Irvine. She has a B.A. in history from the University of California, Irvine, where she graduated Magna Cum Laude and Phi Beta Kappa. Prior to joining the Board on Human-Systems Integration (BOHSI), she worked as a Program Officer with the Board on Atmospheric Sciences and Climate of the National Research Council. She served as study director for *Climate Stabilization Targets: Emissions, Concentrations, and Impacts over Decades to Millennia* and *When Weather Matters: Science and Service to Meet Critical Societal Needs*. During her time with BOHSI, she has served as study director for *The Effects of Commuting on Pilot Fatigue* and *Mine Safety: Essential Components of Self-Escape* as well as provided oversight to *Assessment of Staffing Needs of Systems Specialists in Aviation*. She has nearly a decade's worth of experience as a program manager and community organizer in the fields of public health and youth advocacy in Boston, Massachusetts.